National Parks: Biodiversity, Conservation and Tourism

NATIONAL PARKS: BIODIVERSITY, CONSERVATION AND TOURISM

ANGUS O'REILLY
AND
DORAN MURPHY
EDITORS

Nova Science Publishers, Inc.
New York

Copyright © 2010 by Nova Science Publishers, Inc.

All rights reserved. No part of this book may be reproduced, stored in a retrieval system or transmitted in any form or by any means: electronic, electrostatic, magnetic, tape, mechanical photocopying, recording or otherwise without the written permission of the Publisher.

For permission to use material from this book please contact us:
Telephone 631-231-7269; Fax 631-231-8175
Web Site: http://www.novapublishers.com

NOTICE TO THE READER

The Publisher has taken reasonable care in the preparation of this book, but makes no expressed or implied warranty of any kind and assumes no responsibility for any errors or omissions. No liability is assumed for incidental or consequential damages in connection with or arising out of information contained in this book. The Publisher shall not be liable for any special, consequential, or exemplary damages resulting, in whole or in part, from the readers' use of, or reliance upon, this material. Any parts of this book based on government reports are so indicated and copyright is claimed for those parts to the extent applicable to compilations of such works.

Independent verification should be sought for any data, advice or recommendations contained in this book. In addition, no responsibility is assumed by the publisher for any injury and/or damage to persons or property arising from any methods, products, instructions, ideas or otherwise contained in this publication.

This publication is designed to provide accurate and authoritative information with regard to the subject matter covered herein. It is sold with the clear understanding that the Publisher is not engaged in rendering legal or any other professional services. If legal or any other expert assistance is required, the services of a competent person should be sought. FROM A DECLARATION OF PARTICIPANTS JOINTLY ADOPTED BY A COMMITTEE OF THE AMERICAN BAR ASSOCIATION AND A COMMITTEE OF PUBLISHERS.

LIBRARY OF CONGRESS CATALOGING-IN-PUBLICATION DATA

National parks biodiversity, conservation, and tourism / [edited by] Angus O'Reilly and Doran Murphy.
    p. cm.
  Includes index.
  ISBN 978-1-60741-465-0 (hardcover)
 1. National parks and reserves--Management. 2. Biodiversity conservation. I. O'Reilly, Angus. II. Murphy, Doran.
  SB481.N326 2009
  333.78'3--dc22
                              2009024636

Published by Nova Science Publishers, Inc. ✢ *New York*

CONTENTS

Preface		vii
Chapter 1	The Forgotten Nature of National Parks *Vincent Devictor and Laurent Godet*	1
Chapter 2	Balancing Management Needs for Conserving Biodiversity in Grand Canyon National Park *Jerald C. Mast and Joy N. Mast*	25
Chapter 3	Determination of Optimum Management Strategy for Küre Mountains National Park in Turkey *İsmet Daşdemir and Ersin Güngör*	43
Chapter 4	The Concept of Ecotourism Development with FQFD in the Kinmen National Park *Tsuen-Ho Hsu and Ling-Zhong Lin*	61
Chapter 5	Protection of Nature – Chamois in the Slovak National Parks *Štefančíková Astéria*	85
Chapter 6	The Dilemma of Balancing Conservation and Strong Tourism Interests in a Small National Park: The Case of Amboseli, Kenya *Moses Makonjio Okello, John W. Kiringe and John M. Kioko*	117
Chapter 7	Sacred Groves: Informal Protected Areas in the High Altitudes of Eastern Himalaya, Arunachal Pradesh, Northeast India: Traditional Beliefs, Biodiversity and Conservation *A.R. Barbhuiya, M.L. Khan, A. Arunachalam, S.D. Prabhu and V. Chavan*	131
Chapter 8	High Temperature Environments Represent a Model for the Analysis of Bacterial Dynamics and Preservation of National Parks *M.C. Portillo and J.M. Gonzalez*	147

Chapter 9	Microbial Diversity Supporting Unique Ecosystems within National Parks. The Doñana National Park as an Example *M.C. Portillo, M. Reina, L. Serrano and J.M. Gonzalez*	**161**
Chapter 10	Fire: A Threat and a Lacking Tool for Biodiversity Conservation in Brazilian National Parks *L. Koproski, P.R. Mangini and J.G. Goldammer*	**173**
Chapter 11	Valuation of the Benefits of Wildlife Tourism in Remote Protected Areas: The Case of Gros Morne *Roberto Martínez-Espiñeira*	**181**
Index		**201**

PREFACE

National Parks (NPs) are mainly designed to protect the remaining "wilderness" of a given country and have primarily focused on the conservation of extraordinary areas or emblematic species. National parks have many roles among which "preserving nature" has become a matter of considerable social, political, economical and scientific concern. One of the major problems concerning National Parks is how to preserve their landscapes and biodiversity. While the diversity of plants and animals can be experimentally assessed, their protection involves the maintenance of their ecosystems and periodic monitoring. Any change in an environment can certainly have some effect on the plants and animals living there and so, the consequences of changes at a variety of scales is hard to predict although variations often lead to a reduction or homogenization of animal and plant diversity. In the long run, the positive role played by National Parks for nature conservation and tourism will be maintained if we ensure that social, economic and environmental goals are closely aligned. This new important book gathers the latest research in this field.

National Parks (NPs) are mainly designed to protect the remaining "wilderness" of a given country and have primarily focused on the conservation of extraordinary areas or of emblematic species. However, to halt the current biodiversity crisis, the protection of selected sites dedicated to most threatened species is clearly not sufficient. In Chapter 1, we argue that NPs should now consider a more 'ordinary nature', which has been neglected in conservation and tourism guidelines so far. This ordinary nature encompasses many familiar aspects of the biodiversity, including widespread and abundant species. We first use the French NP network as a case study to highlight why ordinary nature was not considered in this network. We further show how considering common and familiar species could be helpful to track global change impacts on biodiversity, derive original management applications, but also revive the interest of tourism for their 'everyday nature'. In the long run, the positive role played by NPs for nature conservation and tourism will be maintained if we ensure that social, economic and environmental goals are closely aligned. Expanding traditional NP policies to include ordinary nature would be a promising step in that direction.

The National Park Service faces difficult challenges presented by the competing mandates of preserving biodiversity of park ecosystems and providing recreational opportunities for park visitors. Several issues that currently challenge management of the Grand Canyon National Park illustrate this conflict. Much attention has been paid to the struggle of fish native to the Colorado River in the Grand Canyon such as the humpback chub in conditions created by the Glen Canyon Dam. Chapter 2 reviews the issue of managing the riparian habitat along the Colorado River highlighting the Glen Canyon Dam's harmful impacts on native fish, and introduces research on similar impacts on riparian vegetation, particularly on the largest native riparian tree in Grand Canyon National park flora. Impacts of the dam are complex, however, and effective conservation management (which is determined largely outside the Park Service itself) must also take into consideration such positive dam impacts as the emergence of a thriving white-river rafting industry and a world-class trout fishery. These management challenges require sound conservation practices informed by environmental science and decision-making consistent with the progressive values imbedded in the Park Services' mission, as well as an uncomfortable acceptance that, barring dramatic change in societal values, difficult trade-offs may be inevitable.

There are 39 national parks in Turkey and totally 877.771 hectares area is used for this aim. One of them is Küre Mountains National Park (KMNP), which has 37.000 hectares area and was decelerated in the year 2000. It is the first national park which its boundaries determined by contribution and participation of different society groups in Turkey. In this study, it is aimed to determine optimum management strategy for KMNP by turning out preferences of different society groups (local villagers, public institution and nongovernmental organization representatives, potential tourists) and providing their participation to the management process effectively. To realize this aim, it has been studied to develop a management model taking into consideration social and economic values as well as ecological values. It is also investigated a management model such as *to manage conservation areas by participation principle* and *to use income from conservation areas for developing local villagers*.

In Chapter 3, the present situation of the national park is turned out by SWOT analysis, and some alternative management strategies (scenarios) for the national park are developed based on the findings in this stage. The four attributes such as managing type, entrance fee, sharing income and administrative structure which each one has three sub-levels are taken into consideration while developing alternative management scenarios. The nine orthogonal alternative management strategies based on these attributes and their sub-levels are submitted to 462 interviewees selected by layer-random sampling method. The interviewees arranged these strategies according to their preferences. The preference results are evaluated by conjoint analysis, and thus the optimum management strategy is determined for the national park as a managing system with *the balance conservation use, taking US$ 10 entrance fee from visitors, sharing the 70% of income to the national park management and its 30% to local villagers, and administrating with the cooperation of state, nongovernmental organizations and local villagers*. Taking into consideration the optimum management strategy in the management plan of KMNP, which is though preparation, will help the plan to be applicable and dynamic structure, prevent conflicts among local villagers and the national park management, and thus, to contribute to sustainable management of the national park.

Chapter 4 examines local responses to potential ecotourism development in the Kinmen national park located off the southeastern coast of Fujian Province in Xiamen Bay. Ecotourism management has become an economic issue for tourism industry in conservation and development projects. The traditional approaches of biodiversity conservation concerning areas or national parks protection have been considered ineffective and unethical due to the externally imposed rules and regulations on local people. The establishment of national parks and other forms of protected areas in Taiwan in the past has regarded human activities in such areas as incompatible to the goals of natural resource conservation. Conflicts between the authority of national park and indigenous people were reported during the establishment of national parks in Taiwan, which resulted in the relocation and imposition of constraints on indigenous people.

The development and implementation of ecotourism and integrated conservation development projects have been advocated throughout the study to gain local support. The aim of this study is to propose an approach for the issue of indigenous communities and local government ecotourism management based on the framework of Fuzzy Quality Function Deployment (FQFD), a methodology which has been successfully adopted in analyzing human's attitudes and intention. In particular, this paper addresses the issue of how to deploy the house of quality (HoQ) effectively and efficiently toward key dimensions of ecotourism and tourism guideline of the authority. Fuzzy logic is also adopted to deal with the ill-defined nature of the local residents' attitude preference required in the proposed HoQ. The case of Kinmen national park is presented to demonstrate the implementation of the proposed FQFD in ecotourism development. The effective and appropriate management directions for ecotourism development acquired by applying the proposed FQFD, thus, enables the national park authority to achieve an environmental, social, and politico-economic conditions.

Slovak economy is influenced an expressive qualitative and quantitative changes nowadays. It used to be controlled by state exclusively, but at present is transformed into liberal market regime connected with the changes of the ownership and multiple problems to protection and exploitation of natural resources. A lot of plant and animal species disappeared, some became threatened as the result of expansive exploitation of natural resources. From total number of 548 free-living animals 153 are threatened. Tatras chamois (Rupicapra rupicapra tatrica Blahout, 1971) was included as a critically threatened species (IUCN Red List of Threatened Species) (Caprinae Specialist Group 2000). It is also classified as a threatened species by EMA (European Mammal Assessment) (Aulagnier et al. 2008). In the territory of Slovakia , the chamois is native only in High Tatras, West and Belianske Tatras. The fact, that the Tatras chamois are not only the glacial relict, but also the Tatras endemic subspecies (Blahout, 1971), considerably increases their cultural, historic and environmental importance. In order to prevent the chamois extinction, during 1969 – 1976, there was an attempt to create viable population of the Tatras chamois outside their original area. Therefore the Tatras chamois has been introduced to the central part of the Low Tatras National Park. The chamois of Alpine origin from Bohemia and Moravia has been introduced in the 60s into the national parks of Slovak Paradise and Veľká Fatra. The introduction of alpine chamois was hunting motivation of atractive species (Kratochvíl 1981). Chapter 5 analyses data concerning some aspects of the protection of nature in general, but predominantly concentrated on the chamois –their origin, ecological demands in the Slovak national parks, and also analyses the reasons of the depression the Tatras endemic subspecies in detail.

Amboseli is a small park located in the wildlife rich area of Tsavo-Amboseli Ecosystem in Southern Kenya. It is one of the leading parks in terms of absolute tourism revenue generated per year and one of the highest in terms of tourism revenue per unit conservation area. Its revenue is sometimes in excess of Ksh. 100 Million (US$1.33 Million) per year. However, as important as this park is to national and local economy, it has many challenges to its viability in terms of balancing conservation and tourism interests. Several reasons explain this. Its gazettement as a national park in 1974 did not carefully consider implications for the small size in light of its critical role as a dry season wildlife refuge. It is only a fragment of a formerly larger ecosystem, which has now been taken over by human settlement and other incompatible land uses to wildlife conservation. With mainly tourism revenue as a motivation, the government displaced the Maasai from the park and promised to provide them with water from Amboseli swamps or they could water their livestock in the park if this agreement failed. This historical agreement together with rising elephant population, increasing management and tourism infrastructure in the park, the contraction and land use changes in Amboseli Maasai dispersal areas and combine to make Amboseli National Park an increasingly unsustainable conservation unit. Chapter 6 discusses the dilemma of managing Amboseli for biodiversity conservation and as a popular tourist destination.

As presented in Chapter 7, sacred groves are 'traditionally managed' forest patches that functionally link the social life and forest management system of a region. They are the repositories of economical, medicinal, rare, threatened and endemic species and can be regarded as the remnant of the primary forests left untouched/undisturbed by the local inhabitants and protected by local communities due to beliefs that the deities reside in these forests. Arunachal Pradesh, the 'land of rising sun', is located in the northeast region of India, sharing international boundaries with Bhutan, China, Tibet and Myanmar, and is unparalleled in the world for its concentrations, isolation and diversity of tribal cultures and biological diversity. Thus it falls under one of the eight global *mega-diversity hotspots* in the world. It lies between 91°30′–97°30′E longitudes and 26°28′–29°30′N latitudes, covering an area of 83,743 km^2. Approximately 94% of the area is covered by forests, 17.21% of which is very dense, and 45.35% moderately dense. Open areas comprise 18.38% and nonforest areas comprise 18.91%. With a tribal population of about one million represented by 21 major tribal groups with more than 100 ethnically distinct subgroups and over 50 distinct dialects, Arunachal Pradesh contains a good number of sacred groves particularly attached to the Buddhist monasteries, called *gompa forests*, which are managed by the Buddhist community (Monpa and Sherdukpens) of Arunachal Pradesh. These monasteries are mainly found in the West Kameng and Tawang districts of Arunachal Pradesh. Besides these gompa forests, there are a good number of sacred groves in the Tawang and West Kameng districts of Arunachal Pradesh related to the community culture and beliefs. So far, about 63 such sacred groves have been explored as part of a pilot study, including the geographic information, physical and biological attributes and traditional myths associated with the sacred grove. Besides a number of formal protected areas in the region, these informal protected areas, i.e., sacred groves, also play a vital role in the conservation of the significant biodiversity of the region. In the study area, which is more highly valued than the other parts of Arunachal Pradesh, the local Buddhist community also provides tourism services to visitors. The tourism potential in relation to sacred groves and cultural resources is great, but is lacking at present, and needs to be evaluated and appreciated with the active participation of local communities, government bodies and NGOs for the better and sustainable biodiversity services as well as ecotourism. Thus, the rich and enormously diverse biodiversity and cultural heritage of the region may also provide an opportunity for the development of culture tourism in the region. In this regard, the numerous tribes and their multifaceted fairs and festivals can be a powerful attraction to magnetize tourists. An existing scenario of tribal transformation has, however, put pressure on these sacred groves of different sizes. Thus, pro-conservation activities are also admissible for biodiversity management in the Indian eastern Himalayan region and Arunachal Pradesh, northeast India.

As discussed in Chapter 8, several National Parks located in different countries hold high temperature environments such as hot springs or volcanic sites. Among them, Teide National Park and Timanfaya National Park in Canary Islands (Spain), Kamchatka National Park in the Peninsula of Kamchatka (Russia), or Yellowstone National Park (USA) are some well known parks with hot environments. Microbial community studies at these sites have reported the presence of high and low temperature microorganisms which could show critical information on the exposure of these parks to diverse allocthonous microorganisms. Since high temperature environments are prohibitive to mesophilic, temperate microorganisms, the detection of these microorganisms indicates elevated dispersion rates and suggests an incredible colonizing potential. Similar results have been obtained in several National Parks from different world locations corroborating these findings. Since microorganisms play essential roles in biogeochemical cycles of major elements, maintaining their functionality and diversity represent decisive aspects for preserving the equilibrium of their communities and consequently conserve the integrity and variety of environments and landscapes protected within the National Parks.

National Parks represent unique sites in need of preservation for future generations. Although frequently overlooked, the role of microorganisms in the functioning and biogeochemical cycling of elements is essential to support the maintenance of these ecosystems, and consequently their preservations within National Parks. Recent studies have provided serious grounds to confirm that microbial communities present in natural environments are much more diverse than previously imagined. In Chapter 9, we present a case study of a singular environment, the freshwater ponds of the Doñana National Park (Spain). Differences between microbial communities developing in close proximity can be detected suggesting the existence of drastically distinctive niches and microhabitats at a reduced spatial scale. The role of the invisible microorganisms must be considered when managing park conservation and analyzing ecosystem long-term stability. Consequences of the interactive role of microorganisms and environmental and geochemical factors can lead to ecosystem changes with important consequences for nutrient cycling and availability. Preservation of the huge diversity of microbial life in National Parks is a must if the equilibrium and stability of these environments and ecosystems is to be preserved.

As presented in Chapter 10, with more than 1.5 million species, Brazil is recognized as megadiversity country, hosting between 10 and 20% of the species already classified in the world. It is composed of six continental biomes, two of which are known as biodiversity hotspots: the Cerrado and the Atlantic Forest [1]. Nowadays, all Brazilian biomes are modified by human interventions, and their representative biodiversity is endangered at some level. To protect biodiversity in the last decades the number of protected areas in Brazil has grown rapidly. Currently there are nearly 300 federal protected areas officially established, with more than 24,000,000 hectares protected in 63 national parks. Worldwide fires in natural areas cause high environmental losses, negatively impacting global conservation efforts. The situation is not different in Brazil, where wildfires constitute a periodic occurrence and threat to some biomes, jeopardizing ecosystem functioning, biodiversity and atmosphere/climate-related processes. From 1979 to 2005, a total of 2,502 wildfires were officially registered in Federal Brazilian Protected Areas; 1,633 occurred in national parks. In 1979, only three occurrences were registered, contrasting with 196 occurrences in 2005 [2]. Even though these numbers should be higher, because many fire events are not registered or the records are improper, they are sufficient to represent the increasing number of fires in the last decades and to address fire as an important issue to biological diversity conservation in Brazilian National Parks.

In Chapter 11 the travel cost method is used to estimate the demand for wildlife viewing at a national park and to estimate consumer surplus measures associated with the access to the park for visitors mainly attracted to the park by the opportunities to observe wildlife. We use data collected on-site from Gros Morne National Park in Newfoundland&Labrador. Since the data were collected at the park, the analysis takes into account problems related to on-site sampling by using a negative binomial model. This model accounts for the non-negative and integer nature of the dependent variable and also corrects for the truncation and endogenous stratification in its distribution, as well as overdispersion. The results provide relevant insights for the management of recreational areas and for the preservation of biodiversity assets as an economic asset.

In: National Parks: Biodiversity, Conservation and Tourism
Editors: A. O'Reilly and D. Murphy, pp. 1-23

ISBN: 978-1-60741-465-0
© 2010 Nova Science Publishers, Inc.

Chapter 1

THE FORGOTTEN NATURE OF NATIONAL PARKS

Vincent Devictor[a] and Laurent Godet[b]

[a]Edward Grey Institute, Department of Zoology, University of Oxford,
Oxford OX1 3PS, UK ; Tour du Valat, le Sambuc, Arles, France;
and CNRS, UMR 5554, Institut des Sciences de l'Evolution de Montpellier,
Université Montpellier II, Montpellier, France.
[b]CNRS, UMR 7208 BOREA, CRESCO, 38 rue du Port Blanc,
Dinard, France

Abstract

National Parks (NPs) are mainly designed to protect the remaining "wilderness" of a given country and have primarily focused on the conservation of extraordinary areas or of emblematic species. However, to halt the current biodiversity crisis, the protection of selected sites dedicated to most threatened species is clearly not sufficient. In this chapter, we argue that NPs should now consider a more 'ordinary nature', which has been neglected in conservation and tourism guidelines so far. This ordinary nature encompasses many familiar aspects of the biodiversity, including widespread and abundant species. We first use the French NP network as a case study to highlight why ordinary nature was not considered in this network. We further show how considering common and familiar species could be helpful to track global change impacts on biodiversity, derive original management applications, but also revive the interest of tourism for their 'everyday nature'. In the long run, the positive role played by NPs for nature conservation and tourism will be maintained if we ensure that social, economic and environmental goals are closely aligned. Expanding traditional NP policies to include ordinary nature would be a promising step in that direction.

Introduction

NPs have many roles among which 'preserving nature' has become a matter of considerable social, political, economic and scientific concern. Are NPs useful? What are they supposed to do (or not) for conservation and tourism? These questions have been debated for nearly five decades. For instance, early conservationists have investigated how NPs should be designed to maximize species richness: is a single large area doing better than several small? (the so-called 'sloss' debate: single large or several small, Diamond 1975). More recently, such reasoning has been extended to the whole planet with the hotspot concept. The question is, again, how much area is enough to protect the largest number of species? More or less complex tools (including spatial statistics and Geographic Information Systems) were thus proposed to 'optimize' the design of NPs, i.e., to include somehow the greatest amount of the remaining wilderness in the smallest area (Myers et al. 2000).

Interestingly, this stimulating debate is anchored itself in a particular conception of conservation. The key issue of this conception is that protecting taxonomic diversity is important per se, which also lies in the belief that threatened species and rarity are 'more attractive'. Conservation policies of NPs are therefore inevitably biased toward certain species sharing specific attributes (rare, endemic, emblematic, red-listed, threatened, etc.). This conception is well-established, as rare species are also more vulnerable (they are more prone to genetic and demographic stochastic events [Matthies et al. 2004]). Focusing on threatened species also provides a practical basis for management guidelines: listing and monitoring red-listed species have indeed become very popular conservation tools for NPs (Farrier et al. 2007). In this context, the fact that some species are systematically 'forgotten' by NPs seems inevitable, and the need to consider the conservation interest of species that are not critically endangered is rather counterintuitive.

But in the context of the current global changes, should NPs still be solely driven by this biased conception of conservation? In particular, we now expect many species to be driven out of NP boundaries because of climate warming. Moreover, focusing on some species only is unlikely sustainable in the long run: any species depends on biotic interactions and ecosystem processes involving many other common species. The fate of common species therefore inevitably affects directly species of traditional conservation interest (e.g., common species can be the prey of emblematic predators). Consequently, if NPs are changing under global changes one should anticipate whether and how common species will be affected.

In this chapter, we try to go beyond what NPs are currently doing for conservation and tourism to highlight that NPs should now also integrate new conservation issues that has been usually forgotten by traditional conservation plans. In this respect, we use 'ordinary nature' both in an ecological perspective (i.e., abundant and widespread species and habitats) and a social perspective (i.e., familiar species and habitats that belong to 'everyday nature'). This nature contrasts with rare, endangered natural elements and hot spots of endemism and diversity, i.e., 'extraordinary nature'. We also use the term 'unprotected matrix' to designate the non-protected land or marine surface of a given country.

We first use the French NP network to illustrate why, how and where French NPs are currently designed and how ordinary nature is included in this network. We then show that NPs have a key role to play in new conservation issues that encompass ordinary nature. We finally highlight that ordinary nature can also play an important role in tourism in the NPs.

I. 'Extraordinary' and 'Ordinary' Nature in the French National Parks

French NPs have to achieve two missions in nature conservation: protecting extraordinary nature in which humans are absent, and conserving ordinary nature in which humans are present. However, we will show that in practice, conservation efforts are biased: ordinary nature is not considered to be on an equal footing with extraordinary nature in NP policy.

A. NPs: A National Showcase of a Natural Heritage

Protected areas are often considered one of the most effective tools for conserving biodiversity (Possingham et al. 2006). Among the more than 100,000 protected areas, the 3,881 NPs are one kind of protected area that are found all over the world (Shape et al. 2003), and which form an exceptional network of non-perturbed (or little-perturbed) habitats where highly threatened species can be preserved. Some species are even restricted to NPs, like the northern hairy-nosed wombat (*Lasiorhinus krefftii*) (Woolnough & Johnson 2000, in: Possingham et al. 2006). Besides, some NPs were considered to be so exceptional that they were early integrated in the list of the UNESCO World Natural Heritage (e.g., Kilimandjaro NP in Tanzania).

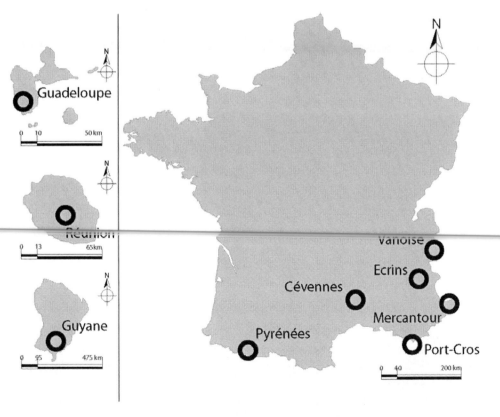

Figure 1. The nine French national parks.

For instance, everyone has heard about the first emblematic NPs of the world: the Yellowstone NP created in 1872 and the Yosemite NP in 1890, both in North America. These first parks had even been created before the total unification of the United States, especially encouraged by the great thinkers such as the painter George Catlin (1796–1872) and the writings of Henry David Thoreau (1817–1862) or John Muir (1838–1914). These first parks were designed to preserve the dramatic and exceptional landscapes of the 'Wild West' and a part of something that has no translation in most languages: the so-called 'wilderness'.

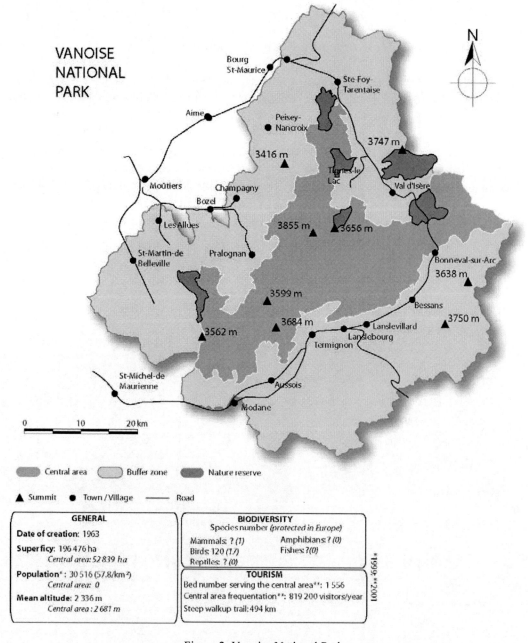

Figure 2. Vanoise National Park.

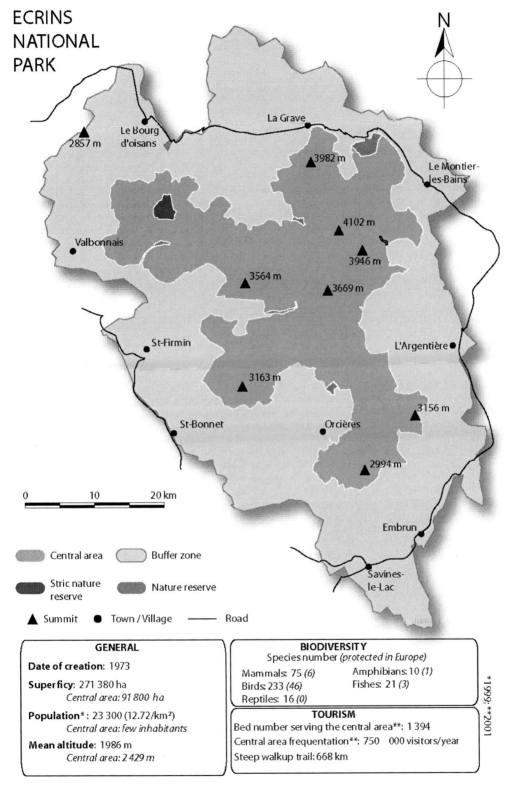

Figure 3. Ecrins National Park.

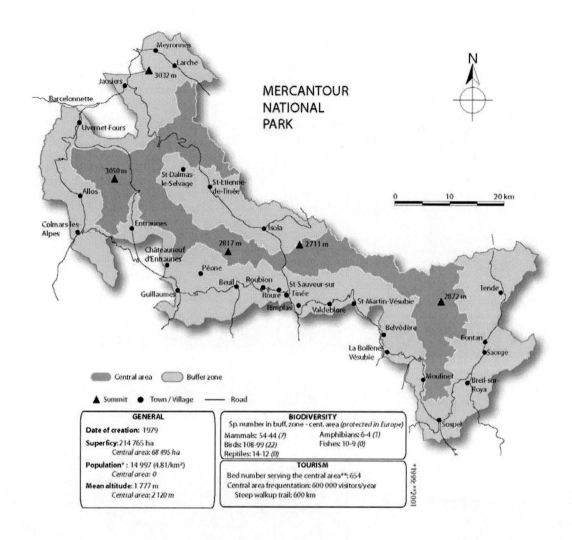

Figure 4. Mercantour National Park.

In France, the nine NPs (Figure 1) are also often considered to delimitate the most emblematic, and among the largest and the wildest protected areas of the national territory. Among the six metropolitan NPs, five are located in mountainous and sparsely populated areas (Pyrénées, Cévennes, Vanoise, Ecrins, Mercantour; see Figures 2–6). They include exceptional landscapes and their associated charismatic species such as big ongulates (chamois *Rupicapra rupicapra*, izard *Rupicapra pyrenaica*, alpine ibex *Capra ibex*), large carnivores (wolf *Canis lupus*, or brown bear *Ursus arctos*) and large raptors (griffon vulture *Gyps fulvus*, cinereous vulture *Aegypius monachus*, golden eagle *Aquila chrysaetos*, etc.). The other metropolitan NP, Port-Cros (Figure 7), is the first European marine NP. Despite the small area covered by this NP (1,993 ha), it is an emblematic protected area of the Mediterranean Sea: it is, for example, the first French Specially Protected Area of Mediterranean Importance (SPAMI). The three additional French NPs are overseas and correspond either to hot spots of endemism (Guadeloupe and Réunion, Figures 8 and 9) or hot spots of species diversity (Guyane, Figure 10). Generally speaking, all of these areas constitute a panel of the national natural wonders.

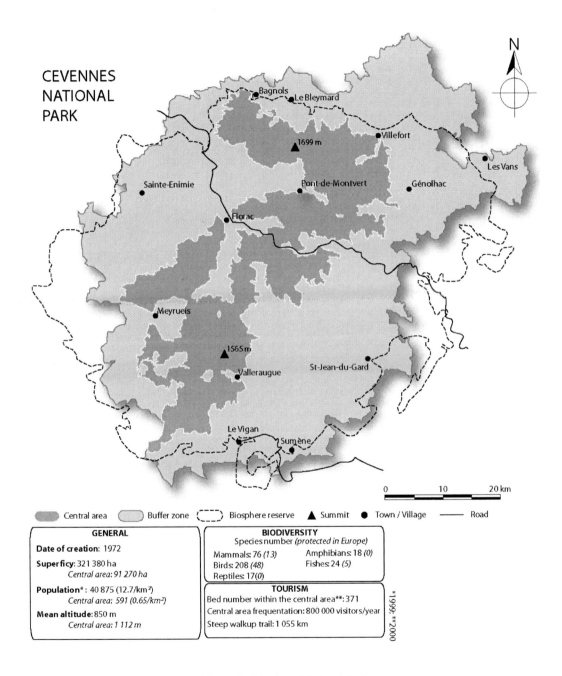

Figure 5. Cévennes National Park.

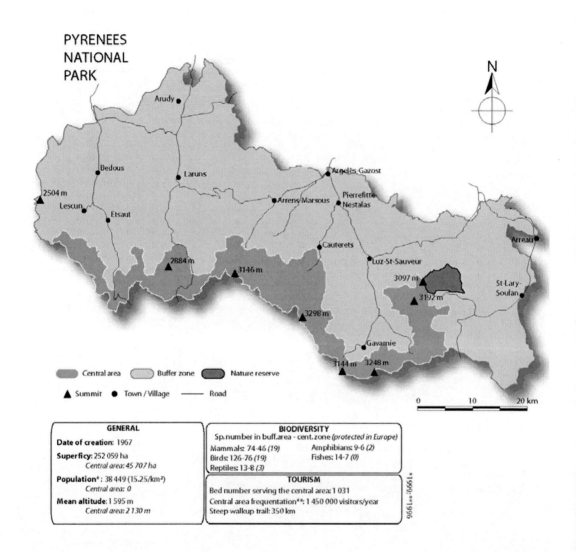

Figure 6. Pyrénées National Park.

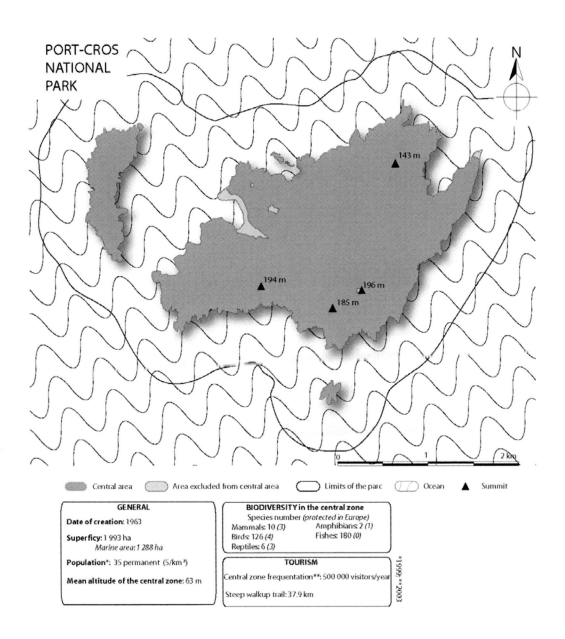

Figure 7. Port-Cros National Park.

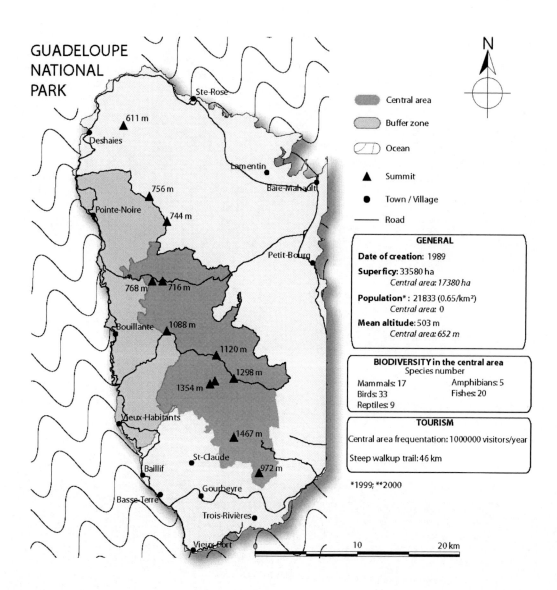

Figure 8. Guadeloupe National Park.

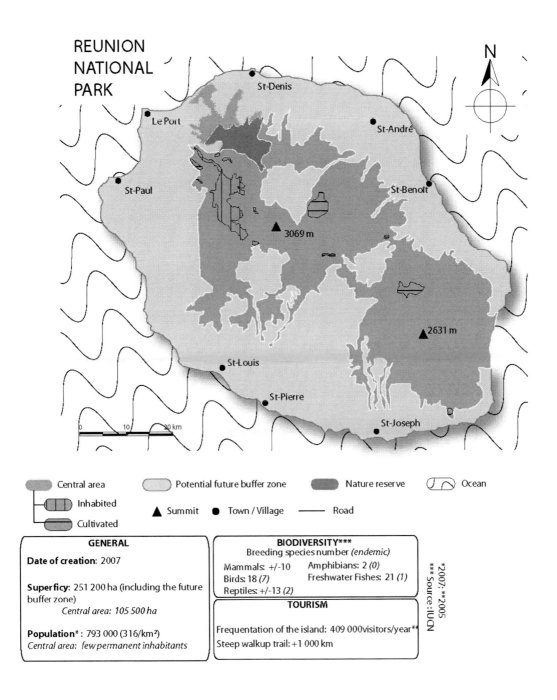

Figure 9. Réunion National Park.

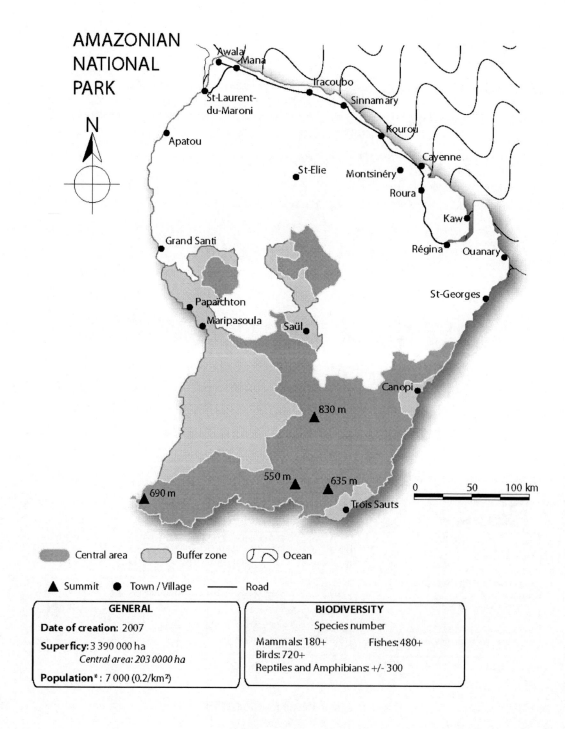

Figure 10. Amazonian (Guyane) National Park.

B. Protecting a 'Wild' Nature while Conserving Anthropo-Ecosystems

Although French NPs share similarities with any other NPs, they also have some specificities. In particular, when the first French NPs have been created, in the 1960s, there was no wilderness left in France. Even the first NP created in 1963 (Vanoise, Figure 2), was implanted around a valley (Maurienne Valley) industrialised since the end of the nineteenth century. Consequently, as soon as the beginning of the twentieth century, French NPs could not include huge uninhabited areas such as in other NPs of the New World.

In fact, opportunities and motivations for creating French NPs were highly context-dependent. Historically, three different park projects emerged no latter than in the 1920s. The first one (1920s) came from the French Alpine Club (FAC). This club wanted to create a park to protect the Izard *R. rupricaria*, in order to be able to observe this emblematic species in the future. But, interestingly, the FAC was opposed to the conservation of large carnivorous (Wolf *C. lupus*, or Brown bear *U. arctos*), considered as pests. The second one (1930s), was a project coming from hunter associations to create a small park, contiguous to the Italian NP 'Grand Paradisio', to protect the Alpine ibex *C. ibex*, in order to be able to hunt this species in the future (Couturier 1943). Finally, in the 1950s, the emergence of the first NP was also due to the strong and long lobbying work of a former mayor, Gilbert André, an outsider with high political connections which soon became an influential figure in the founding of the National Park (Thompson 1999).

Gilbert André, French politicians, journalists, but also intellectuals then became aware of the necessity of the creation of a NP in the French Alps. According to Gilbert André, this modern park had to be a 'cultural park' where traditional activities had to be maintained. Based on these pioneer ideas, the French administration decided to create a new protection tool: the French NP, coupling the spirit of these three different projects (preserving nature, allowing hunting, and acknowledging the importance of cultural heritage). In doing so, all the different participants involved in the negotiation of the creation of a NP could be satisfied (see Mauz 2002). In consequence, schematically, the Vanoise NP, and all the following NPs of the French territory were then constituted similarly and composed of two concentric areas:

i. a **central zone** where human activities are strictly controlled (scientific activities are often the only tolerated activities), and where strict nature reserves can be set up. Except for the Cévennes NP, central areas are uninhabited (or with few permanent inhabitants), and they mainly concern high altitude areas. The central zone corresponds to the IUCN Category II: 'NP: protected area managed for ecosystem *protection* and recreation' (IUCN 1994). With the strict nature reserves, this is the highest protection level in France.
ii. a **buffer zone**, where human activities and nature have to coexist in harmony. In this area the development has to be exemplary. The buffer zone corresponds to the IUCN Category V: 'Protected landscape/seascape: protected area managed mainly for landscape/seascape *conservation* and recreation' (IUCN 1994).

The French NPs are therefore deeply rooted in this compromise between strictly protected areas (i.e., the central zone) *versus* conserved areas with a human development in inhabited areas (i.e., the buffer zone). Note that although French NPs are among the largest protected areas in the country, they are small compared to NP of Africa, Australia or America. They have been created especially where the human development was not too important (although not where human activities were absent). They are also very recent protection measures (the oldest being 45 years old and the two last ones only two years old). Besides, their creation is so recent that the parks were implemented in areas where the large charismatic species had already disappeared or were very scarce. For instance, in the 1950s, the Alpine ibex *C. ibex*, was very scarce in the French Alps, the wolf *C. lupus* was already absent, and only 50 brown bears *U. arctos* survived in the French Pyrénées (Maurin 1992).

Consequently, French NPs were early considered not only as being important for protecting an exceptional biodiversity: they also should eventually allow the *restoration* or the *reintroduction of* the populations of large birds and mammals. Nowadays, most of these reintroduced or restored populations are quite wealthy (except the brown bear) and are the success showcase of the NPs (for example, vultures were successfully reintroduced in the Cévennes, and the Alpine ibex *C. ibex* in alpine parks).

Although it cannot be the spearhead of their communication policy, note that French NPs (as other protected areas in France) are implicitly supposed to monitor and preserve the ordinary nature. Indeed, NP managers generally act in favour of nature considered as a whole, at least in the buffer zone. The NPs are also active participants of the common nature monitoring programs such as the French Breeding Bird Survey, the National Observatory of the Meadows (Vanoise), etc. Some NPs are also pilot sites for ambitious projects towards the conservation of biodiversity in its largest extent (e.g., the comprehensive biodiversity inventory project in the Mercantour NP, bringing together more than 300 specialists from all over the world to collect data on many threatened and unthreatened species).

However, it is still the exceptional landscapes and the charismatic species which steals the limelight in the NPs. The need to monitor the ordinary nature seems to be a hardly used argument in the communication policy of the NPs. This situation is quite paradoxical if we consider that the National park includes a large amount of ordinary nature: the buffer zones represent more than 86% of the total superficies of the NPs (4,255,782 ha of buffer zones, 676,991 ha of central areas). These zones are inhabited, and delimitate areas where people live and work and including the so-called 'everyday nature'.

In fact, the more or less absence of the ordinary nature in the NP communication policy can be explained by the general wisdom assuming that these area have an exceptional role to play towards an exceptional nature. In consequence, one believes that the public opinion won't judge their action conducted for common species despite their potential regression in the national territory in the last years. NPs are rather more assessed by the public opinion and the politics on their ability to save emblematic species. In other words, NPs managers have to conserve more a *natural heritage* than a *biodiversity*. They don't have to protect only fauna and flora species or habitats, they have to protect a nature that the public feel important to be protected (a nature one can be proud of) and wants to pass down from generation to generation. But is this situation only true for terrestrial National parks? One may indeed ask whether this situation is different for marine biodiversity conservation.

C. Are French National Parks Suitable to Protect Marine Diversity?

Surprisingly, the marine nature is almost totally absent from the French NPs. Yet, at a global scale, marine protected areas cover around half a percent of the world's seas (Roberts & Hawkins 2000). In France, the only marine area included in a NP is in Port-Cros, which only covers 1,288 ha of marine area (Figure 7). Moreover, this is the only NP without any buffer zone. Strangely enough, the insular Guadeloupe and Réunion NPs are restricted to a strictly terrestrial perimeter (Figures 8 and 9). How can we explain this quasi absence of the marine environment in the French NPs?

Firstly, the marine environment is much less-known than the terrestrial one. For example, in France, the first descriptions of the marine benthic communities occurring in the French-side of the English Channel are only 40 years old (Cabioch 1968, Cabioch & al. 1978, Retière 1979). Moreover, while there are complete databases of terrestrial habitat mapping (such as land cover layers) over large spatial scales, such databases hardly exist for marine environment. The first reason explaining the mismatch between conservation needs and National Park design for marine areas is therefore technical: it is simply too difficult to conserve areas without precise maps or recent biological data.

Secondly, including an area in a NP (and especially a central zone of a NP) requires the identification of particular characteristics to justify the need of such protection planning. In the marine environment, there are much less areas recognized as worth being protected. Most marine areas of conservation interest are seen as 'small islands' over a large unknown ocean. In this respect, coral reefs and hot spots of diversity or endemism are the only parts of the ocean which benefit from large marine protected areas (Godet et al. 2008). For example, the largest marine protected areas include the Great Barrier Reef Marine Park, the Galapagos NP and the Northwestern Hawaiian Islands Marine National Monument.

But the European seas do not have such large exceptional features: no large coral reef barriers, no abyssal zones, no kelp forests, etc. Moreover, the beauty of the European underwater seascapes (such as rocks covered by Pink sea-fans *Eunicella verrucosa*, or Jewel anemones *Corynactis viridis*) are only accessible to few people such as divers, and they have never been largely popularized through the media. It is also difficult to find exceptional species to protect: there is no endemic invertebrate marine species along the French coasts (Boucher 1992). In other words, France but also many other European states may have some difficulties to select marine areas which may become a national showcase of something of conservation interest, neither at a seascape level nor at a species level. Nowadays the only marine benthic habitats benefiting from a legal protection are 'structured habitats' such as eelgrass beds (*Zostera marina* and *Zostera noltii* beds, *Posidonia oceanica* beds, etc.) or maerl *Lithothamnion* spp. beds. These habitats cover very restricted areas (for example, the largest French *Z. marina* beds cover approximately 500 ha), and can therefore only benefit from local protection measures.

Thirdly, the marine ordinary nature is drastically different from the terrestrial one. Terrestrial ordinary nature is inhabited by human societies and humans have modeled the landscapes according to their own spatial scale. On the contrary, marine ordinary nature is not inhabited by humans and the spatial scales of the energy and material fluxes largely exceed the spatial scale humans are used to experiment. Moreover, seascapes are not visible, and their natural boundaries are much fuzzier than in terrestrial systems (except for the 'structured habitats': seagrass meadows, biogenic reefs, etc.). In consequence, the marine ordinary nature is not an 'everyday nature'. People cannot sentimentally appropriate the marine nature like the terrestrial one. Moreover, the public involvement in the conservation of marine ordinary nature is much less developed. This issue has been remarkably highlighted by Richmond (1994):

> "Most people are familiar with terrestrial habitats and can relate to a walk in the woods. Few, however, have experienced the wonders of a coral reef except for occasionally viewing Jacques Cousteau special. Whilst it is easy to capture images of rain forests being cut down and to collect data to quantify the magnitude of habitat destruction on land, it is more difficult to study and to document coral reef processes and degradation."

and by Norse & Crowder (2005):

> "Even more than poverty, affluence, technology, and greed, it is ignorance and indifference that are the enemies of marine biodiversity."

These differences between terrestrial and marine nature probably explain the quasi-exclusion of the marine nature from the NPs. The European marine nature neither corresponds to the exceptional nature standard of a NP central zone nor to the traditional ordinary nature of a NP buffer area. The originalities of the marine environment have even lead France to develop a new kind a protected area in 2006: the marine natural park, managed by a much less restrictive policy than in terrestrial NPs.

Using the French NP network as a case study, we thus showed that both terrestrial and marine NPs were clearly not primarily designed for ordinary nature. Yet, we believe that the exclusion of marine nature from NPs, but also the lack of communication about the terrestrial ordinary nature must be overcome. Although rarity, species diversity and exceptional characteristics of a landscape can be useful criteria to select the important areas which can be considered as a showcase of a national natural heritage, NPs also have to be a conservation model. For this particular reason, they have to be able to conserve an ordinary nature which represents increasing conservation stakes both for the biodiversity itself but also for tourism.

II. Considering Ordinary Nature in NPs: A Matter of Conservation Interest?

One may ask why NPs should care for the conservation of common and widespread species. The first answer could be "why not?" Indeed, there is growing evidence that protecting biodiversity is more a question of protecting biotic interactions and processes rather than species *per se*. The other answers come from new conservation and scientific issues raised by recent global changes.

A. NPs under Global Changes: Ordinary Nature and New Conservation Issues

Many common species are now showing a large spatial and/or temporal large-scale decline (e.g., marsupials, Fisher et al. 2003; birds, Julliard et al. 2004; amphibians, Stuart et al. 2004; plants, Smart et al. 2005; insects, Conrad et al. 2006; butterflies, Wenzel et al. 2006). The estimates of local population losses of common species are much higher than those of species extinctions (Hughes et al. 1997), and many common species are likely accumulating a debt, i.e., a future ecological cost of current habitat destruction (Tilman et al. 2002). However, the need to pay more attention to severe depletion of common species has been hardly highlighted or only recently (Gaston & Fuller 2008). Surprisingly, considering the potential role played by NPs for conservation using ordinary nature has also hardly been investigated (but see Devictor et al. 2007).

This decline in common species populations may be attributed to many human pressures occurring at large scale which have been documented throughout the last decade: the loss of grasslands, healthy forests and wetlands, and other critical habitats from multiple environmental threats such as urban sprawl, energy development, and the spread of industrialized agriculture occurring in human-dominated landscapes. The consequences of these profound land-use changes on common species are now complicated by their responses to rapid climate warming (McClanahan et al. 2008).

The decrease in common species most likely results in the loss of good candidates to ensure future responses and adaptation to global changes. Population diversity provides insurance against change in environmental conditions and confers greater flexibility on individuals for coping with anthropogenic or non-anthropogenic stress (Luck et al. 2003). Genetic variability is also often higher in common species (Cole 2003) so that protecting genetic diversity of widespread and common species was shown to be necessary to prevent negative consequences of habitat fragmentation (Honnay & Jacquemyn 2006).

Moreover, ecological services based on the presence of large populations (e.g., for production, regulation, pollination, dispersion) are principally altered by the local loss of populations of common species. For instance, the decline of common bird species populations in NPs would lead to the decline of many dispersers, raptors, predators and pollinators, which would threaten vital ecosystem processes and functions (Şekercioğlu 2006). Thus, although a species can be widespread and abundant in a NP or/and in the unprotected matrix, its local functional role can be substantially reduced under certain population sizes, and become 'functionally extinct' (Şekercioğlu 2006; Gaston & Fuller 2008).

Moreover, although they only cover a small proportion of the land-surface compared to the unprotected matrix, NPs often include important habitats for many migratory species. These species may need to halt in more hospitable places and hence benefit from a direct refugee effect provided by NPs. More generally, the spatial distribution of many species is now expected to change in response to climate warming (McClanahan et al. 2008) so that NPs but also other protected areas may buffer the negative effects of landscape degradation occurring in the unprotected matrix. In the long-run, NPs may thus act as permanent or temporary local source of individuals for many common species.

B. Building Relevant Indicators Using the Fate of Common Species in NPs

Acknowledging the importance of ordinary nature should not lead to the identification of more and more unrealistic conservation goals for NPs, but should rather contribute to underline what NPs can presently easily achieve towards more holistic conservation approaches. The first role NP could play for conservation of ordinary nature is simply to monitor common species to produce useful biodiversity barometers of common species' fate. Indeed, the best way to study the consequences of global changes on biodiversity is to use monitoring programs yielding considerable amounts of standardized data across taxa and spatial scales. Common species thus provide biodiversity indicators both scientifically sound and useful for decision makers (Gregory et al. 2005).

Common species can also serve as proxy for measuring complex patterns or processes of conservation interest. For instance, spatial patterns of species richness of a given group is often better described by recording distributions of common rather than rare species (Lennon et al. 2004). Studying relative abundance of widespread species is also a powerful scientific tool to test general ecological predictions about species responses to land use-changes (Devictor et al. 2008), global warming (Julliard et al. 2004) or to more specific management actions (e.g., impacts of forest harvesting, Pohl et al. 2007). Therefore, comparing trends of common species within and outside NPs could provide critical information about how biodiversity react to land-use and climate changes.

More specifically, comparing the long-term trend of a particular common species (or set of species) within and outside NPs would be a direct and inexpensive way to assess whether the trend is similar, better, or worth when measured in NP than the same trend measured at larger spatial scale (Devictor et al. 2007). This method could be further used to assess the effectiveness of a specific management option in the NP: if the trend changes following a management action while the reference trend (e.g., measured outside the NP) remains constant, it would provide a natural experimental assessment of the management action.

But perhaps more important than these scientific considerations is the potential role played by NPs on tourism. The year 2003 was the first in which more people lived in urban areas than rural ones. By 2050 as many as three quarters of the world's human population will live in cities and suburbs (Cohn 2005). Humans may have progressively fewer opportunities for first-hand experience of other life-forms. However, such contacts may be vital in stimulating an appreciation of the natural world and the desire to conserve it (Collar 2003; Pyle 2003). Therefore, one should investigate whether the ordinary nature can also be useful for tourism in NPs.

III. Improving Tourism in NPs Based on Ordinary Nature

The importance of common species for NPs is also raised by recent societal issues. Indeed, 'familiar species', 'wider countryside', 'ordinary nature', and 'everyday nature', are terms now frequently used in conservation biology and land use policy. Nature protection is no longer solely considered as set apart from human activities and restricted to emblematic or rare species. The protection of a social nature encompassing a variety of environments and cultural contexts has gained credence (Kaplan et al. 1999). Therefore, it is also likely that focusing on ordinary nature could be helpful to develop relevant tourism programs in NPs.

A. Tourists are not Only Interested in Extraordinary Nature

NPs have to fulfil the double function of public welcome and biodiversity protection. First, we can wonder why it is important to exclude and control human activities from specific areas on the one hand, and to encourage the tourism activity in the same areas on the other hand. At first glance, these goals are indeed very different or even opposed, and this duality may have disastrous consequences. Indeed, the NP administration has been sometimes too permissive with local communities by tolerating human activities within the NP perimeters (e.g., alpine NPs did not prevent the creation of ski runs within the NPs). Moreover, some NPs even directly favour the development of human activities. For instance, the Amazonian (Guyane) NP is very new, and is also restricted to the southern part of the department, where human activities are the less developed. But surprisingly, one of the NP projects is to develop tourism, especially by making rivers more navigable, by opening tourist information centre, and by teaching autochthonous people tourism jobs (http://www.parc-guyane.gf/site.php?id=7%22).

In this context, one may think that for biodiversity conservation, maintaining low human population density, limited number of economic activities and absence of access roads were probably better protection tools than such a modern 'protection' project increasing local human pressure. It seems indeed paradoxical to open these exceptional areas to public visitors especially when these areas were initially 'naturally' preserved. In fact, it would have probably been also illogical to recognize the exceptional characteristics of this area and its natural heritage value, and to forbid its access. This restriction would lead to forbid the enjoyment of the nature, precisely based on the grounds of why we like it.

Given this duality, we could usefully distinguish two kinds of extraordinary nature: a rare and vulnerable nature on the one hand, and an extraordinary nature on the sense of an *exceptionnal* nature (but not threatened) on the other hand (emblematic, charismatic species, dramatic and beautiful landscapes, etc.). Encouraging the access to the first one can be dangerous for rare and threatened species. Encouraging the access to the second one is justified from an ethical and pedagogical point of view. Moreover, it is likely that most tourists are precisely looking for this second part of nature in the NPs. In other words, for the great majority of tourists, it is certainly more exciting to watch a large ungulates' family in a short distance in the dramatic mountainous landscapes of the Vanoise NP than observing the very rare coleoptera *Danosoma fasciata* on a rotten trunk of the same NP.

NP managers have to deal with this dichotomy to manage a subtle harmony in which ordinary can be very helpful. Indeed, while rare and threatened species or habitats may still be considered as priority (and their access strongly controlled), managers could also emphasize the existence of the extraordinary-side of the ordinary nature. Clearly, this point of view is based on highly subjective criteria, different from 'objective' criteria such as rarity or vulnerability levels. This is a representation, passing through the perception filter of everyone. Nature would thus appear to be extraordinary for some people (tourists living in big cities for example) and ordinary for others (conservationists for example). But for NP managers, the ordinary nature, non-threatened and common species, can all form an interesting pedagogic basis for tourism purpose.

B. Developing a More Conscious Tourism using Ordinary Nature Protection

Using the fate of common species in NPs should be a straightforward way to emphasize conservation matters that are regionally and locally meaningful for people and at the same time globally responsive. Indeed, people who live in urbanized environments often have a great appreciation of many common species, such as birds (McKinney 2002) or biodiversity in general (Fischer and Young 2007). People most often interact with most common species and thus should feel more concerned by the fate of these species.

If the ordinary nature is showed and explained to tourists in NPs, people should realize that they will meet similar species and habitats where they live and work. In this respect, showing and explaining that all species interact should be encouraged. For instance, one should explain that abundant common plants are highly needed to preserve ungulates which are themselves crucial to preserve charismatic raptors. Similarly, the beauty of common plant species could be emphasized rather than only the rarity of endemic species. Highlighting the importance of the ordinary nature should, in the long run, increase the number of people participating in the integration of biodiversity in their everyday life when they leave NPs.

Moreover, most people still rely on a close knowledge of common species, or attend to 'a familiar nature' in the context of outdoor recreation. This link has created several cultural connections between people and common species through individual perceptions, beliefs and values. Therefore, most familiar species may contribute to a person's attachment to a particular place, become part of a person's identity, and therefore support an individual's psychological wellbeing (Horwitz et al. 2001). These value judgments are now recognized as essential instruments for an improved design and communication of biodiversity policies (Miller 2005; Fischer & Young 2007; D'elia et al. 2008). These ordinary aspects of NPs should thus also be considered to find public support for biodiversity conservation in NPs.

Conclusion

Many conservationists have calculated that, ideally, most of the existing species on Earth could be maintained by protecting only 12% of the land surface. In fact, this reasoning is particularly static and unconvincing in solving the problem of the current biodiversity crisis (Rodrigues et al. 2004). The main reason is that, in essence, nature cannot be protected by being put into a box: nature is everywhere, is moving in space and time and is evolving. This dynamic aspect of biodiversity pleads for greater consideration of ordinary nature.

Obviously, the fact that NPs should not forget ordinary nature must neither result in any exaggeration leading to a domestic view of NPs nor mask other conservation issues. Forecasting what NPs can also do for (and with) common species must not unprofitably polarize conservation planning in different approaches against each other. The key idea is rather to emphasize an existing, but overlooked, continuum from commonness to rarity and from "ordinary" to "extraordinary" nature. In this respect, considering the fate of both common and rare species simultaneously can be highly relevant. For instance, genetic patterns from common species might be used to designate geographic areas where both common and rare species show evidence of historic isolation, thus increasing the conservation significance of those regions (Whiteley et al. 2006). Similarly, comparing the trends of common and rare species can be useful to identify critical life history traits that will influence species responses to particular management regimes (Farnsworth 2007). These approaches can shed light on areas or populations (within or outside NPs) of particular interest that would be neglected if common species are not considered. Although conservation biology was early considered as 'a science of scarcity and diversity' (Soulé 1986), enlarging conservation issues to common species is now highly needed. All of these changes challenge conservationists to consider how more common and familiar valuation aspects of biodiversity can be included in NP policies

References

Boucher, P. (1992). *Terre Océane*. Paris, Imprimerie Nationale.

Cabioch, L. (1968). Contribution à la connaissance des peuplements benthique de la Manche occidentale. *Cahier de Biologie Marine*, **9**, 493-720.

Cabioch, L ; Gentil F. ; Glaçon R. & Retière, C. (1978). Le bassin occidental de la Manche, modèle de distribution de peuplements benthiques dans une mer à fortes marées. *Journal de la Recherche Océanographique*, **3**, 249.

Cohn, J.P. (2005). Urban wildlife. *BioScience*, **55**, 201-205.

Cole, C.T. (2003). Genetic variation in rare and common plants. *Annual Review of Ecology and Systematics*, **34**, 213-237.

Collar, N.J. (2003). Beyond value: biodiversity and the freedom of the mind. *Global Ecology and Biogeography*, **12**, 265-269.

Conrad, K. F.; Warren, M. S.; Fox, R.; Parsons, M. S.; Woiwod, I. P. (2006). Rapid declines of common, widespread British moths provide evidence of an insect biodiversity crisis. *Biological Conservation*, **132**, 279-291.

Couturier, M. A. (1943). Projet d'un parc national à bouquetins en France. *Revue de Géographie alpine*, **31**, 39-398.

D'elia, J.; Zwartjes, M. & McCarthy, S. (2008). Considering legal viability and societal values when deciding what to conserve under the U.S. Endangered Species Act. *Conservation Biology*, **22**, 1072-1074.

Devictor, V.; Godet, L.; Julliard, R.; Couvet, D. & Jiguet, F. (2007). Can common species benefit from protected areas? *Biological Conservation*, **139**, 29-36.

Devictor, V.; Julliard, R.; Clavel, J.; Jiguet, F.; Lee, A. & Couvet, D. (2008) Functional biotic homogenization of bird communities in disturbed landscapes. *Global Ecology and Biogeography*, **17**, 252-261.

Diamond, J. M. (1975). The Island Dilemma: Lessons of Modern Biogeographic Studies for the Design of Natural Reserves. *Biological Conservation*, **7**, 129-146.

Farnsworth, E. J. (2007). Plant life history traits of rare versus frequent plant taxa of sandplains: Implications for research and management trials. *Biological Conservation*, **136**, 44-52.

Farrier, D.; Whelan, R. & Mooney, C. (2007). Threatened species listing as a trigger for conservation action. *Environmental Science & Policy*, **10**, 219-229.

Fisher, D. O.; Blomberg, S. P. & Owens, I. P. F. (2003). Extrinsic versus intrinsic factors in the decline and extinction of Australian marsupials. *Proceedings of the Royal Society B*, **270**, 1801-1808.

Fischer, A. & Young, J. C. (2007). Understanding mental constructs of biodiversity: Implications for biodiversity management and conservation. *Biological Conservation*, **136**, 271-282.

Gaston, K. J. & Fuller, R. A. (2008). Commonness, population depletion, and conservation biology. *Trends in Ecology & Evolution*, **23**, 14-19.

Godet, L.; Toupoint, N.; Olivier, F.; Fournier, J. & Retière, C. (2008). Considering the functional value of common marine species as a conservation stake: the case of the sandmason worm Lanice conchilega beds. *Ambio*, **37**, 347-355.

Gregory, R. D.; van Strien, A.; Vorisek, P.; Meyling, A. W. G.; Noble, D. G.; Foppen, R. P. B. & Gibbons, D. W. (2005). Developping indicators for European birds. *Proceedings of the Royal Society B*, **360**, 269-288.

Honnay, O. & Jacquemyn, H. (2006). Susceptibility of common and rare plant species to the genetic consequences of habitat fragmentation. *Conservation Biology*, **21**, 823-831.

Horwitz, P.; Lindsay, M. & O'Connor, M. (2001). Biodiversity, endemism, sense of place, and public health: inter-relationships for Australian inland aquatic systems. *Ecosystem Health*, **7**, 253-265.

Hughes, J. B.; Daily, G. C. & Ehrlich, P. R. (1997). Population diversity: its extent and extinction. *Science*, **278**, 689-692.

IUCN (1994). *Guidelines for protected areas management categories* (IUCN). Switzerland, Cambridge, UK and Gland.

Julliard, R.; Jiguet, F. & Couvet, D. (2004). Common birds facing global changes: what makes a species at risk? *Global Change Biology*, **10**, 148-154.

Kaplan, R.; Ryan, R. L. & Kaplan, S. (1999). *With People in Mind: Design and Management for Everyday Nature*. Washington DC, Island press.

Lennon, J. J.; Koleff, P.; Greenwood, J. J. D. & Gaston, K. J. (2004). Contribution of rarity and commonness to patterns of species richness. *Ecology Letters*, **7**, 81-87.

Luck, G. W.; Daily, G. C. & Ehrlich, P. R. (2003). Population diversity and ecosystem services. *Trends in Ecology & Evolution*, **18**, 331-336.

Matthies, D.; Bräuer, I.; Maibom, W.; Tscharntke, T. (2004). Population size and the risk of local extinction: empirical evidence from rare plants. *Oikos*, **105**, 481-488.

Maurin, H. (Dir.) (1992). Inventaire de la faune de France. Paris, Muséum National d'Histoire Naturelle, Nathan.

Mauz, I. (2002). Comment est née la conception française des parcs nationaux? *Revue de Géographie Alpine*, **10**, 33-44.

McClanahan, T.R., Cinner, J.E., Maina, J., Graham, N.A.J., Daw, T.M., Stead, S.M., Wamukota, A., Brown, K., Ateweberhan, M., Venus, V. & Polunin, N.V.C. (2008) Conservation action in a changing climate. *Conservation Letters* **1**, 53-59.

McKinney, M. L. (2002). Urbanization, biodiversity and conservation. *BioScience*, **52**, 883-890.

Miller, J. R. (2005). Biodiversity conservation and the extinction of experience. *Trends in Ecology & Evolution*, **20**, 430-434.

Myers, N; Mittermeier, R. A.; Mittermeier, C. G.; da Fonseca, G. A. B. & Kent, J. (2000). Biodiversity hotspots for conservation priorities. *Nature*, **403**, 853–858.

Norse, E. A. & Crowder L. B. (2005). *Marine Conservation Biology*. Washington, Island Press.

Pohl, G. R.; Langor, D. W. & Spence, J. R. (2007). Rove beetles and ground beetles (*Coleoptera*: *Staphylinidae*, *Carabidae*) as indicators of harvest and regeneration practices in western Canadian foothills forests. *Biological Conservation*, **137**, 294-307.

Possingham, H. P.; Wilson, K.A.; Andelman, S. J. & Vynne, C. H. (2006). Protected Areas: Goals, Limitations, and Design. In M. J. Groom; G. K. Meefe & C. R. Carroll, Editors

(Eds.), *Principles of Conservation Biology* (3rd edition, pp. 509-533). Sunderland MA. Sinauer Associates.

Pyle, R. M. (2003). Nature matrix: reconnecting people and nature. *Oryx,* **37**, 206-214.

Retière, C. (1979). Contribution à la connaissance des peuplements benthiques du Golfe Normano-Breton. Thèse de doctorat, 431pp.

Richmond, R. H. (1994). Coral reefs: Pollution impacts. *Forum for Applied Research and Public Policy,* **9**, 54-57.

Roberts, C. M. & Hawkins, J. P. (2000). Fully protected marine reserves: a guide. WWF Endangered Seas Campaign. Washington, DC (USA), and University of York, York (UK). Available from: www.panda.org/resources/publications/water/mpreserves/mar_dwnld.htm

Rodrigues, A. S. L.; Andelman, S. J.; Bakarr, M. I.; Boitani, L.; Brooks, T. M.; Cowling, R. M.; Fishpool, L. D. C.; Fonseca, G. A. B.; Gaston, K. J.; Hoffmann, M.; Long, J. S.; Marquet, P. A.; Pilgrim, J. D.; Pressey, R. L.; Schipper, J.; Sechrest, W.; Stuart, S. N.; Underhill, L. G.; Waller, R. W.; Watts, M. E. J. & Yan, X. (2004). Effectiveness of the global protected area network in representing species diversity. *Nature,* **428**, 640-643.

Şekercioğlu, Ç. (2006) Increasing awareness of avian ecological function. *Trends in Ecology & Evolution,* **221**, 464-471.

Shape, S.; Blyth, S.; Fish, L.; Fox, P. & Spalding M. (2003). The United Nations List of Protected Areas. IUCN, Gland, Switzerland and Cambridge, UK and UNEP-WCMC, Cambridge, UK.

Smart, S. M.; Bunce, R. G. H.; Marrs, R.; LeDuc, M.; Firbank, L. G.; Maskell, L. C.; Scott, W. A.; Thompson, K. & Walker, K. J. (2005). Large-scale changes in the abundance of common higher plant species across Britain between 1978, 1990 and 1998 as a consequence of human activity: Tests of hypothesised changes in trait representation. *Biological Conservation,* **124**, 355-371.

Soulé, M. E. (1986). Conservation biology: the science of scarcity and diversity. Sinauer, Sunderland.

Stuart, S. N.; Chanson, J. S.; Cox, N. A.; Young, B. E.; Rodrigues, A. S. L.; Fischman, D. L. & Waller, R.W. (2004). Status and trends of amphibian declines and extinctions worldwide. *Science* **306**, 1783-1786.

Thompson, I. B. (1999). Sustainable rural development in the context of a high mountain national park: The parc national de la Vanoise, France. *Scottish Geographical Journal* **115**, 297-318.

Tilman, D.; May, R. M.; Lehman, C. L. & Nowak, M. A. (2002). Habitat destruction and the extinction debt. *Nature,* **371**, 65-66.

Wenzel, M.; Schmitt, T.; Weitzel, M. & Seitz, A. (2006). The severe decline of butterflies on western German calcareous grasslands during the last 30 years: a conservation problem. *Biological Conservation,* **128**, 542-552.

Whiteley, A.; Spruell, P. & Allendorf, F. W. (2006). Can common species provide valuable information for conservation? *Molecular Ecology,* **15**, 2767-2786.

In: National Parks: Biodiversity, Conservation and Tourism
Editors: A. O'Reilly and D. Murphy, pp. 25-42
ISBN: 978-1-60741-465-0
© 2010 Nova Science Publishers, Inc.

Chapter 2

BALANCING MANAGEMENT NEEDS FOR CONSERVING BIODIVERSITY IN GRAND CANYON NATIONAL PARK

Jerald C. Mast and Joy N. Mast
Carthage College

Abstract

The National Park Service faces difficult challenges presented by the competing mandates of preserving biodiversity of park ecosystems and providing recreational opportunities for park visitors. Several issues that currently challenge management of the Grand Canyon National Park illustrate this conflict. Much attention has been paid to the struggle of fish native to the Colorado River in the Grand Canyon such as the humpback chub in conditions created by the Glen Canyon Dam. We review the issue of managing the riparian habitat along the Colorado River highlighting the Glen Canyon Dam's harmful impacts on native fish, and introduce research on similar impacts on riparian vegetation, particularly on the largest native riparian tree in Grand Canyon National park flora. Impacts of the dam are complex, however, and effective conservation management (which is determined largely outside the Park Service itself) must also take into consideration such positive dam impacts as the emergence of a thriving white-river rafting industry and a world-class trout fishery. These management challenges require sound conservation practices informed by environmental science and decision-making consistent with the progressive values imbedded in the Park Services' mission, as well as an uncomfortable acceptance that, barring dramatic change in societal values, difficult trade-offs may be inevitable.

Introduction

American writer John Lawson Stoddard (1850-1931) wrote of the Grand Canyon, "To stand upon the edge of this stupendous gorge, as it receives its earliest greeting from the god of day, is to enjoy in a moment compensation for long years of ordinary, uneventful life" (1905). While the exact proportion of the Grand Canyon's four million annual visitors who would agree with Stoddard's claim is uncertain, that so many of us continue to come and return is testimony enough to the profound love people have for the Grand Canyon and of the restorative effects it has upon us.

The first national parks were created out of a sense of national aestheticism, and the Grand Canyon, established as a National Monument by Theodore Roosevelt in 1908, and elevated to National Park status by Congress in 1919, epitomized this impulse to preserve scenic resources. Its majesty and grandeur simultaneously captivate the imagination while instilling in the observer a profound sense of humility. This effect on the viewer has helped make the Grand Canyon the most iconic of landscapes in the topography of American culture. The original national park philosophy of monumentalism described by Alfred Runte (1987) grounds the value of the parks in their unique and compelling scenic attributes, and in the psychological effects these attributes have upon us collectively and individually. The preeminence of the Grand Canyon as an example of a unique visual monument worthy of preservation is ably described by Theodore Roosevelt's famous statement in 1903 that "…the Grand Canyon…[is]…a natural wonder which, so far as I know, is in kind absolutely unparalleled throughout the rest of the world…Leave it as it is. You cannot improve upon on it, and man can only mar it. What you can do is to keep it for your children, your children's children, and for all who come after you, as the one great sight every American should see" (Lewis, 1906, p. 327).

Roosevelt's preservationist charge was both prescient and paradoxical. Visitation rates provide ample empirical evidence to support his claim that the Grand Canyon possessed and would retain a central importance to the American nation as a source of wonder and pride. Yet his directive to "leave it as it is" has not been followed, and the consequences of human impacts on the Grand Canyon illustrate two types of threat to the values preservationism seeks to protect: first, an external threat that originates from management decisions informed by a separate philosophy of resource usage (conservationism), and second, an internal tension within preservationism created by the desires to preserve a place and experience it (Nash, 1982). The victory of American preservationism depended on two critical factors: its willingness to limit its ambitions by ceding territory to conservationist purposes where the commodity values were high, and the popularity of its cause where preservationists did choose to make a stand. Dangers to the Park System exist within both factors. On the one hand, vulnerability comes in the form of ecosystems that extend beyond the Park Service's administrative boundaries, and on the other is the specter of loving wilderness to death.

Roosevelt's paradoxical political sentiments to both use the Grand Canyon and preserve it were given the force of law at the National Park Service's inception. The Park Service's Organic Act stipulates the Service's mission is to "promote and regulate the use of Federal areas known as national parks, monuments, and reservations…[and] conserve the scenery and the natural and historic objects and the wildlife therein and to provide for the enjoyment of future generations." (16 U.S.C. § 1) Thus its mission allows for both a recreational mandate and preservation mandate. While it is true that these competing imperatives are narrower in scope than the more conflictual purposes of the multiple-use agencies like the National Forest Service or the Bureau of Land Management, these competing mandates place the Park Service in the position of making political choices about management alternatives. Despite the need to make political choices from time to time, the relatively more focused mission has allowed the Park Service to acquire a less controversial reputation than its more put-upon bureaucratic brethren (Clarke and McCool, 1985). The extractive decisions of these other, conservation-oriented agencies can have a negative impact on the ability of the Park Service to fulfill its mission. Moreover, conflict over the allocation of resources within the Park Service generally reflects inherent discord between use-oriented policies that facilitate recreational opportunities and preservation-oriented policies that sustainably promote the natural integrity of the parks.

Several historical developments in the latter half of the 20th century have served to intensify this discord. First, economic growth created a middle class with enough disposable income to recreate in the out of doors, and the interstate highway system facilitated access to previously remote areas. Thus outdoor recreation grew dramatically in the 1950's and 1960's. Annual visitation rates at the Grand Canyon from 1920-1949 averaged 234,790 visitors; the average annual number of visitors in the 1950's was 899,409 and in the 1960's the annual average was 1,641,894 visitors (National Park Service, 2009). These figures reflect increasing use of and thus impact on the environments of which the National Park system is composed. Second, the development of increasingly sophisticated environmental sciences over the 20th century has resulted in our improved understanding of ecosystem functions generally and the consequences to them of these increasing rates of human activity.

These historical developments resulted in new ecological concerns competing with cultural and recreational interests in the debate over appropriate directions for management of national parks. Increasing scientific research on ecosystems, particularly on the part of the federal land management agencies, intensified starting in the 1960's (Hayes, 2000). A focus on habitat, and related subjects such as aquatic toxicology and biological succession, became an increasingly important trend in environmental science. While the National Park Service has a history of resisting the use of science for ecosystem management in favor of traditional recreational - park management approaches (Parsons, 2004), where science was embraced within the National Park system, it was generally employed to identify the pre-European character of park ecosystems, and to restore them to those conditions. Where once preservation as a concept had largely aesthetic meaning, it has increasingly acquired biological meaning as well. The policy consequences of increased ecological concern about the well-being of native species harmed by incompatible land-use policies ensure that this science will continue to emerge in a contentious political context.

This chapter reviews the impacts of Glen Canyon Dam on recreation on the Colorado River and on the riparian ecosystems of Grand Canyon National Park and discusses a case study of those same impacts on a native plant species. This discussion will illustrate how the ability of the Park Service to maintain or restore biodiversity can be impeded by policies implemented by other agencies such as the Bureau of Reclamation that affect NPS resources, and complicated by competing park interests such as recreation. The case study focuses on one part of a larger ecological concern for the Grand Canyon National Park - alterations to the riparian environment from upstream and downstream dams on the Colorado River. Minimizing harmful ecological impacts to biodiversity in the riparian environment along the Colorado River through simulated floods involves trade-offs in the commercial benefits created by the dams, but also effects important recreational interests. It is therefore an excellent example of the Park management conundrum of conflict between preservation and use imperatives, but also conflict within preservation objectives themselves.

Dam Construction on the Colorado River

The era of big dam construction in the American west reflected natural resource management consistent with the ethic of progressive conservationism (Hayes, 1999). This perspective typically defines the environment as composed of natural resources to be harnessed for the greater public good, and assumes maximizing the efficiency with which they are used should be the guiding criteria in decision-making. Dam construction in the western United States provided jobs during the Great Depression, a steady supply of electricity for rapid urban development, flood control, and facilitated water storage for desert-based agriculture and the growing population of a thirsty region. Employing science and technology to realize the potential of our natural resources, the marriage of human ingenuity and the environment was seen as a hallmark of national progress. It would be several decades before serious consideration of environmental damage from dam construction and operation would take significant root in the public's consciousness. When it did, concern for harmful effects on the parks would play a central role, for while progressive conservationism that justified the construction of these dams did not predominate within the boundaries of the national parks themselves, it certainly impacted the ecosystems within them.

West of Grand Canyon National Park, the Hoover Dam was created in part to control flooding in the Imperial Valley of California. Beginning in 1932, the Colorado River was diverted around the Hoover Dam building site. In 1933 the first concrete was poured for the Hoover Dam, with water impounding behind the dam into the Black Canyon starting in 1935. By 1941, the Hoover Dam had created Lake Mead, which is 120 miles long, up to 580 feet deep, 1,220 feet in elevation, and covers about 157,900 acres. Hoover Dam generates an average of 4 billion kilowatt-hours of electricity each year, serving 1.3 million persons in California, Nevada and Arizona (Bureau of Reclamation, 2009). Between 1935 and 1963, about 91,000 acre-feet of sediment was deposited in Lake Mead every year.

In part to extend the longevity of Hoover Dam, Glen Canyon Dam was installed ca. 370 miles upstream which helped trap sediments as well as to store water. The storage of water behind Glen Canyon Dam allows the states along the upper basin of the Colorado River (Colorado, Utah, New Mexico, and part of Arizona) to ensure their ability to deliver a legally determined amount of water to states along the lower basin (California, Nevada and most of Arizona) and to Mexico (Adler, 2007). The Glen Canyon Dam buried Glen Canyon beneath the waters of newly created Lake Powell, which at 266 square miles is one of the largest reservoirs in North America. The Glen Canyon Dam is capable of generating about 5,000 GWh (billion watt hours) of hydro-power annually, for 1.7 million end-use customers, most of whom reside over the six state southwestern region (National Research Council, 1996).

The commercial bias of progressive conservationism was manifest in the legislation that created the Glen Canyon Dam and has guided the Bureau of Reclamation's management of it. The Colorado River Storage Act of 1956, which set construction of the dam into motion, stipulated operating rules for the dam that did not allow for the valuing of species or habitat harmed at the expense of the commercial values of water storage and power generation - which were to occur at the greatest level practicable within the bounds of existing water laws (Adler, 2007). While environmental concerns about dams emerged in force by mid-century, a compromise to preserve Dinosaur National Monument from inundation meant construction of the Glen Canyon Dam in 1963 went largely unopposed.

Dam Effects on Recreation along the Colorado River

In addition to the clearly significant commercial benefits created by Hoover and Glen Canyon Dams, the reservoirs they create possess dramatic recreational benefits. Administered by the National Park Service as part of Lake Mead National Recreation Area, Lake Mead averaged nearly 8 million visitors per year from 2000 to 2008, mainly for boating, water skiing, fishing and swimming (National Park Service Public Use Statistics Office, 2009). Lake Mead and its surrounding area constitute the first national recreation area in the United States. Since 2000, approximately 2 million people have visited the Glen Canyon National Recreation Area to recreate on or around Lake Powell annually (National Park Service Public Use Statistics Office, 2009). The Glen Canyon Dam has created conditions facilitating the development of a Blue-Ribbon trout fishery between the dam and Lee's Ferry which receives over 20 thousand annual visitors and generates several million dollars in economic benefits (Slaughter et al., 2003).

Within the Grand Canyon itself, the Glen Canyon Dam has played an important role in the emergence over the last four decades of a thriving whitewater rafting industry on the Colorado River. First successfully run by Major John Wesley Powell in 1869, the Colorado River saw few boaters, mainly government surveyors and a handful of private adventurers, for most of the next century. The first commercial raft trips began in 1938, when Norman Nevills began to take tourists down the river; nevertheless, by 1950, only about 345 persons had run the river through the Grand Canyon (Ghiglieri and Meyers, 2001). The larger trend of outdoor recreation and National Park visitation in the 1950's and 1960's was reflected in increased Colorado River traffic and, aided by the advent of rubber raft technology capable of withstanding the pounding of the river's famed rapids, commercial river-running became significant presence on the river by the 1970's.

Today, it is a major industry, and demand for access to the Colorado River in the Grand Canyon for rafting purposes far exceeds what the Park Service is willing to allow. Sixteen commercial rafting companies hold permits from the National Park Service to operate on the river. By the mid 1990's, 15,000 to 20,000 persons were running the river annually on commercial or private trips, the former generating $23 million in local economic activity (National Research Council, 1996). In 2001, comparable figures were found by economists at Northern Arizona University: approximately 22,000 persons rafted the Colorado River through Grand Canyon National Park that year, generating $21,100,000 of economic output and creating of 357 full-time jobs for the regional economy (Hjerpe and Kim, 2007). However, less than 50% of the expenditures are captured by the local economy, and many of the jobs created are low-wage. Low-wage job creation in regions with higher than average unemployment is nonetheless a positive development. The 2006 National Park Management Policies explicitly call for Park Service relations with local and regional actors for purposes of economic development, which provides an opportunity for an NPS role in addressing both economic leakage and low-wage aspects of the rafting industry (Hjerpe and Kim, 2007).

As noted previously, the Park Service limits the number of people allowed on the river well below the number who would otherwise use it. Currently, the Park Service's Colorado River Management Plan allows for 17,606 persons to traverse the Grand Canyon on commercial trips, and 7,051 persons on private trips, or 24,657 total park visitors running the river annually (National Park Service, 2006). The Park Service regulates these boating experiences for purposes of safety, over-crowding on the river, and to minimize ecological impact along the river corridor.

The Park Service has not always been willingly accommodating of the commercial rafting industry. In 1976, the Park administrators at Grand Canyon introduced a plan to designate the Colorado River corridor as wilderness, which would preclude the use of motorized rafts through the park. Because motorized rafts navigate the 250 mile river section of the park much faster than oar-powered boats, rafting companies can sell many more seats on such boats over the course of the season, making this mode of transportation more convenient for and more popular with the public and more profitable for the industry. The rafting industry's interests were successfully defended by Utah Senator Orrin Hatch, who attached a rider to an Interior Department appropriation bill that prevented the Park Service's plan to remove motorized boats from the river in 1981 (Morehouse, 1996). National Park officials at the Grand Canyon have periodically raised the issue of wilderness designation and the phasing out of motorized rafts in the river corridor since, but so far without significant success.

The Glen Canyon Dam has in many ways facilitated the development of the rafting industry in the Grand Canyon as it is currently constituted. The reduction of seasonal flow fluctuations has opened up river accessibility year-round, whereas once the torrential spring-melt fed floods were not navigable, and low-flow seasons heighten the danger of the white-water experiences through the larger rapids (National Research Council, 1996). The post-dam predictability of the river makes possible the longer-term planning conducive to healthy business. With regards to the alternative dam management plans studied by the Glen Canyon Environmental Studies (discussed later), commercial boatmen preferred plans that reduced the daily fluctuations of the river. These plans allow for more and better access to beach sites for camping purposes, as the higher flows may increase crowding along the river corridor as beach space becomes inundated (National Research Council, 1996).

In their review of fatalities within the Grand Canyon, Ghiglieri and Meyers (2001) contend that the most dangerous component of white-water rafting the Colorado may be the constant cold temperatures created by the dam's releasing of water from the bottom of the Lake Powell. As will be discussed below, one of the current management proposals to benefit native fish habitat is to divert water from the upper depths of the reservoir, which would lower the average temperature of the down-stream river. This might make the river safer by reducing the probability of hypothermia in those who fall in, as well as create likely benefits for endangered native fish species.

Dam Effects on Biodiversity along the Colorado River

The commercial and recreational benefits created by Glen Canyon and Hoover Dams are dramatic, but come at the cost of altering the riparian habitat of the Colorado River downstream of them, thus illustrating a central aspect of the vulnerability of national park ecosystems. Specifically, they must deal with impacts to the park ecosystems from threats external to their jurisdictions, regardless of how effective park managers are at resolving the tensions between use-oriented recreational values and preservation-oriented ecosystem integrity values within the boundaries of the parks themselves (Freemuth, 1991).

The Colorado River flows for 277 miles through Grand Canyon National Park, officially starting at Lees Ferry (river mile 0), 15 miles downstream from Glen Canyon Dam, and ending below Grand Wash Cliffs at the beginning of Lake Mead (river mile 277) (Stevens, 1983). Glen Canyon Dam alters the Colorado River's downstream flow, reducing the natural seasonal variation of flowage and imposing a fluctuating daily cycle based on electricity demands (Howard and Dolan, 1981). Ecological effects of this river regulation include flood reduction, lower mean annual maximum flows, higher median flows, colder and less variable river temperatures, sediment trapping behind the dam up-river, and beach erosion (Thomas et. al., 1960; Carothers and Dolan, 1982). These ecological changes to the Grand Canyon's habitat result in altered spawning habitat for native aquatic species, diminished beach environment for native plants, and reduced sediment loads for seed establishment. At the western end of the Grand Canyon at the beginning of Lake Mead, the environment is radically different with the former canyon under water and new sediment build up providing new habitat for plant establishment. These ecological alterations increase pressure from non-native competitors like rainbow trout (Oncorhynchus mykiss) and tamarisk (Tamarix ramosissima, also commonly called salt cedar) who benefit from the changes.

In 1978, the U.S. Fish and Wildlife Service (FWS) issued a biological opinion that the humpback chub (Gila cypha) and the Colorado pikeminnow (Ptychocheilus lucius, since extirpated) were both endangered by conditions that result from operations of the Glen Canyon Dam. The razorback sucker (Xyrauchen texanus) was listed as endangered in 1991. These fish evolved in pre-dam river conditions characterized by small daily fluctuations in water flow, with very large seasonal fluctuations. They depend on warmer, backwater and near-shore habitats that develop around beaches and sandbars that have been eroded away by the scouring effect of steady, daily fluctuations in post-dam, cold water releases devoid of sediment.

The Endangered Species Act (ESA) of 1973 forbids land owners and land managers such as federal agencies from engaging in actions that "take" individuals of species that are determined by the FWS to be threatened or endangered (16 U.S.C. § 1538 (1)(a) (b)). Federal courts have interpreted this to include actions that harm the habitat on which endangered species depend (Babbitt, Secretary of Interior et. al. v. Sweet Home Chapter of Communities for a Great Oregon et. al. 515 U.S. 687, 1995). Federal agencies must consult with the FWS to ensure their actions do not harm the viability of species listed as endangered or threatened under the ESA.

The famous court case Tennessee Valley Authority vs. Hill illustrated the potential power of the ESA to elevate the interests of endangered fish over commercial values associated with dams (437 U.S. 153, 1978). In the case of the Colorado River in Grand Canyon, the federal statutes authorizing the Glen Canyon Dam and establishing the parameters of its management predate the ESA. These laws governing how the Colorado River is managed include the Colorado Compact, the original treaty between the basin states allocating water allotments between states in the region which, in combination with its subsequent, associated laws, rules, contracts and court cases, form the "Law of the River" (Adler, 2007). The Colorado River Storage Project Act of 1956 and the Colorado River Basin Project Act of 1968 establish Glen Canyon management and as such are part of the Law of the River, and both are clear in their prioritization of values associated with commercial usage.

As noted above, federal agencies like the Bureau of Reclamation are required to consult with the FWS for purposes of endangered species protection, and to avoid actions which would impede the restoration of them. But federal agencies, including the FWS, have interpreted these requirements as applying to agency actions over which the agencies have discretion. Federal courts have consistently reinforced this point in cases in which environmentalists have sought to alter dam management to protect endangered species (see for example, Southwest Center for Biological Diversity v. Bureau of Reclamation [143 F.3d 515, 1997], Defenders of Wildlife v. Norton [257 F. Supp. 2d 53, 2003], and National Association of Home Builders v. Defenders of Wildlife [551 U.S. 644, 2007]). Although federal agencies like the Bureau of Reclamation are bound to adopt actions that protect endangered species and avoid actions that harm them, and consult with the FWS about these duties, federal case law currently holds that agencies' actions that are mandatory as per other law are precluded from these duties under the ESA. Thus the conflicting imperatives of species preservation and commercial utilization greatly complicates legal framework of managing the Glen Canyon Dam.

Dams can operated in ways that approximate natural river conditions to potentially benefit native species, but the cost of these management options is reduced power generation capacity (which cannot occur during very high releases to simulate large floods). This trade-off of lost power generated for more natural river conditions may be seen as a desirable policy goal consistent with the preservation mandate of the Park Service's Organic Act, and the goals established the ESA, but the law governing how the Bureau of Reclamation operates the dam precluded it. This contradiction in legal imperatives led to Congressional intervention in 1992 in the form of the Grand Canyon Protection Act. This policy attempted to balance the conflicting goals of ecological restoration and preservation on the one hand, with the commercial interests of pre-existing dam operations. The unfortunate result, however, was more confusion.

While the Grand Canyon Protection Act stipulates that the dam shall be managed "in such a manner as to protect, mitigate adverse impacts to, and improve the values for which the Grand Canyon National Park and the Glen Canyon National Recreation Area were established", it also maintains that the Secretary of the Interior shall do so "in a manner fully consistent with and subject to" the Colorado River Compact and other aspects of the Law of the River such as the 1956 and 1968 laws (cited by Adler, 2007, p. 145). Congress equivocated, rhetorically committing to both ecosystem restoration along the Colorado River corridor in the Grand Canyon, and commercial exploitation of the water resource created by the Glen Canyon Dam, ignoring the inherent contradictions in these goals.

Uncovering these contradictions and attempting to resolve them depends on information about the environmental consequences of the Glen Canyon Dam in the Grand Canyon. Concern for native species and their habitat motivated and is informed by scientific research. Such research also plays a critical role in the articulation of potential recovery plans, and given the apparent degrees of incompatibility between species protection and other natural resource usage, scientific data plays an important part in the political and legal conflicts involved with managing the Glen Canyon dam. The National Environmental Policy Act (NEPA), passed in 1969, requires federal agencies to publicize proposed actions that may significantly affect the environment, and to study and assess those possible effects. In 1982, the Reagan administration's interest in increasing the electrical generation capacity at the Glen Canyon Dam triggered the NEPA requirements and the Bureau of Reclamation initiated the Glen Canyon Environmental Studies (Powell, 2008). Originally the Glen Canyon Environmental Studies was limited in status to an environmental assessment rather than a full environmental impact statement, for an assessment is narrower in scope, would probe less deeply into environmental impacts and would not require public input. Despite its narrower scope, and to the apparent disappointment of top officials in the Interior Department, the environmental assessment found that the current operations of the Glen Canyon Dam had been adversely affecting wildlife and recreation since the dam had begun operations (Powell, 2008).

By the end of the 1980's new proposals for marketing energy generated by the dam intensified political conflict. After reviewing the Glen Canyon Environmental Studies' environmental assessment, the National Academy of Science endorsed its conclusion that environmental and recreation interests were being harmed by the dam, prompting President George H.W. Bush's Interior Secretary Manuel Lujan to order a full-blown environmental impact statement (Powell, 2008). A primary outcome of these research efforts was the identification of possible consequences of a range of proposed dam management alternatives. The spectrum of studied alternatives ranged from increasing the commercial output of the dam's energy production, to operating the dam in a way that mimicked natural flow patterns as closely as possible.

The Bureau of Reclamation's final environmental impact statement, issued in 1995, rejected the alternative recommended by environmentalists and the FWS that would mimic natural river flows ("seasonally adjusted steady flows") and instead recommended a strategy of dam operation known as "modified low fluctuating flow" or MLFF. The MLFF option minimizes impact on power generation by keeping in place pre-existing annual and monthly flow targets, while trying to reduce impacts on habitat and recreational interests by reducing the range of fluctuation that occurs hourly and daily. It also allows for periodic experimental flood simulations for ecological restoration and recreation purposes. The Bureau of Reclamation's choice was made over objections of the FWS, which the year previously had issued a formal statement (called a Biological Opinion) that the MLFF management strategy would jeopardize the welfare of the endangered humpback chub.

The ESA requires federal agencies to consult with the FWS in order to ensure that their actions do not negatively affect endangered species or their habitat. If an agency chooses to engage in an action the FWS concludes will harm an endangered species, the FWS must issue a plan consisting of "reasonably and prudent" steps that will mitigate the harm caused by the agency's choice of action. In response to Reclamation's preference for MLFF management, the FWS recommended: 1) creation of an adaptive management program in which stakeholders could provide ongoing feedback for purposes of recommending policy corrections; 2) a series of simulated flood experiments where high steady flows were created in the spring and low steady flows created in the summer and fall; 3) selective withdrawal of warm water near the surface of Lake Powell to raise the temperature of the Colorado River downstream from the dam; 4) establish a humpback chub management program for the Little Colorado River; and 5) establish a second spawning colony of humpback chub below the dam (in addition to the colony at the confluence of the Colorado and Little Colorado Rivers).

The Bureau of Reclamation's recommendation of MLFF and the "reasonably and prudent" steps drafted by the FWS were formalized as law in a Record of Decision reached in cooperation with multiple stakeholders and signed by Interior Secretary Bruce Babbitt in 1996. The Record of Decision established the Glen Canyon Adaptive Management Program run by the Adaptive Management Work Group, a twenty-five member body composed of agency, tribal and various stakeholder representatives that advise Interior Secretaries on implementation issues. Progress on other elements of the FWS's "reasonably and prudent" steps began immediately.

In the spring of 1996, the Bureau of Reclamation released 45,000 cubic feet per second (cfs) of Colorado River water for a week to conduct an experimental flood to simulate natural flood conditions, hoping to recharge sandbars and beaches along the river corridor and improve resources for camping, and habitat in terms of approximating pre-dam conditions, while minimizing impacts on endangered species (Stevens et. al., 2001). The experiment was in ways a political success for ecosystem restoration in that it established a precedent in the United States of altering dam management to recreate pre-dam ecosystem processes. In diverting water around the turbines, the experiment represented an official acknowledgement of tangible environmental values, to be considered (at least temporarily) along side more conventional commercial values. Yet with regards to its management objectives, the results were more mixed.

Because the Glen Canyon Dam traps sediment upriver, habitat below the dam now depends on sediment transported by tributary streams and rivers. When the 1996 flood experiment was conducted, there was insufficient sand in the river bed for the flood to distribute. As a result, increases in sandbars and beaches were small and temporary, and eventually eroded to less than pre-dam sizes. More positively, the simulated flood experiment successfully avoided serious harm to several notable endangered species that have adapted to post-dam conditions within the riparian corridor. The Kanab ambersnail (Oxyloma haydeni kanabensis), the peregrine falcon (Falco peregrines), the southwestern willow flycatcher (Empidonax traillii extimus) and other species of concern were either minimally and temporarily affected by the flood, or not affected at all (Stevens et. al., 2001). That some important native species have benefited from post-dam conditions illustrates the complexity of the situation, and the importance of ongoing scientific research and adaptive management approaches.

A second flood simulation took place in 2004, and was timed when large amounts of sediment were being deposited by Colorado River tributaries – resulting in significant increased sandbars through Marble Canyon, but more erosion farther downstream. The Bureau of Reclamation has also implemented mechanical removal of rainbow trout which successfully compete with native fish like the humpback chub and razorback sucker from the Colorado River in the Grand Canyon National Park. Trout have become established in the River, thriving in the cold waters created by the conventional dam operations, and a Blue-Ribbon trout fishery now exists between Lees Ferry and the Dam itself, attracting thousands of sport fishermen annually. The Bureau of Reclamation has successfully established a spawning population of humpback chub on the Little Colorado River and is also working on system that will allow the dam to draw warmer water closer to the surface of Lake Powell to raise temperatures in the Colorado River below the dam to benefit the chub and other native species.

These efforts may be having a positive effect, as the United States Geological Survey, responsible for monitoring research within the Grand Canyon, reports that humpback chub populations have increased by 50% since 2001, however, they still remain lower than 1995 levels (United States Geological Survey, 2008). The rise in population prompted the FWS to reverse its previous position issued in 1994 that the MLFF management option jeopardizes the humpback chub, a decision that sparked protest from environmentalists, and led to a lawsuit discussed below.

In the spring of 2008 a third flood simulation was conducted which illustrated differences between criteria influencing management decisions, as well as tensions between agencies making them. The Bureau of Reclamation conducted a single 60 hour long, 41,500 cfs flood, timing it to take advantage of high sediment loads being delivered by the Paria River (United States Geological Survey, 2008). It followed up the flood by returning to the normal dam operations during the summer, with a series of steady flow releases in the fall. The Bureau of Reclamation plan drew heated criticism when it was announced from the Superintendent of Grand Canyon National Park Steve Martin on several counts: the timing of the subsequent steady flows in the fall represents the elevation of commercial power interests over ecological ones (which would suggest steady flows earlier during the summer when electricity prices are at their highest); a lack of any further flood-level releases in the next four or five years; and a failure to adequately involve the Grand Canyon National Park personnel in planning the experiment (Wilson, 2008).

Controversy over the 2008 experimental flood carried over into federal courts. In December of 2007 the Grand Canyon Trust, an environmental organization and member of the Glen Canyon Adaptive Management Working Group, filed suit against the Bureau of Reclamation and the FWS claiming that Reclamation's preferred MLFF jeopardizes the humpback chub and damages its critical habitat. The Grand Canyon Trust furthered accused the federal agencies of failing to adhere to procedural requirements of the NEPA, and ESA, and the Grand Canyon Protection Act.

In May 2009, a federal district court for Arizona issued several troubling statements about federal operations of the Glen Canyon Dam (Grand Canyon Trust v. Bureau of Reclamation, --- F. Supp. 2D --- WL 1458784 D. Ariz.). It found that the 2008 Biological Opinion issued by the FWS that MLFF management option did not jeopardize the humpback chub to be insufficiently justified with supporting evidence. As such, the court ordered the FWS to revise its 2008 Biological Opinion, and expressed skepticism about the extent to which MLFF was consistent with the ESA given the amount of scientific research that fluctuating flows harmed backwater and near-shore habitat critical to the humpback chub. The choices between managing for the interests of native fish habitat and managing for power generation are becoming clearer. But to frame the conflict as one between fish and power is to fail to account for other ecological values that may be at risk. The next section discusses research into the effects of the Glen Canyon Dam on the largest native riparian tree in the Grand Canyon, the Goodding willow (Salix gooddingii).

Case-Study of Dam Effects on Goodding Willow

Riparian vegetation communities consist of old high water and new high water bands that parallel the river and are based on the different flood lines before and after Glen Canyon Dam construction. For the majority of Grand Canyon National Park, old high water vegetation consists of mesquite (Prosopis glandulosa) and catclaw (Acacia greggii) while new high water zone includes tamarisk, mesquite, seep-willows (Baccharis spp.) and coyote willows (Salix exigua). Before dam construction, frequent flood events limited the amounts of vegetation along the Colorado River in Grand Canyon National Park (Clover and Jotter, 1944). Since the construction of Glen Canyon Dam in 1963, riparian vegetation has expanded and diversified (Carothers and Aitchison, 1976; Purcherelli, 1986; Stevens and Waring, 1988).

Glen Canyon Dam has altered the Colorado River riparian vegetation primarily due to flow regime. Pre-dam river flowage at Lee's Ferry averaged 16,800 cfs, but possessed a large seasonal range, with highs in June in excess of 100,000 cfs and lows in fall and winter as little as 2,500 cfs. Daily fluctuation was much smaller, though occasionally significant. After the reservoir was filled in 1980, the average annual flows remained approximated pre-dam levels, but the seasonal flow range was greatly reduced, while the daily fluctuation in river flowage was increased to a range of as low as1,000 cfs to highs of 31,000 cfs (National Research Council, 1996). These ranges have largely been dictated by the fluctuating needs for power generation, which is generally higher during the day than at night, and higher in summer and winter than in spring or fall.

Our research highlights the Glen Canyon dam's impacts on native riparian vegetation, focusing on the Goodding willow which is the largest native riparian tree in Grand Canyon National park flora (Turner and Karpiscak, 1980). Although a dominant riparian tree species prior to the construction of the dam, Goodding willow is currently at risk due to habitat destruction, competition with introduced species such as tamarisk, and river regulation. Willow recruitment and establishment are dependent on periodic floods (Reichenbacher, 1984).

For our study, we estimated the health of the Goodding willow population (Mast and Waring, 1996; Mast, 1997). Tree ring patterns illustrate river regulation impacts versus climatic influences on establishment and growth. Demographic field studies also help in analyses of willow responses to flood flows. Dendrochronological evidence of willow establishment and health is compared to Colorado River flow records, particularly the impact of Glen Canyon Dam in 1963 and the record post-dam flood in 1983 (recorded at 92,600 cubic feet per second). Age, size, and health data on Goodding willows throughout the river corridor in the Grand Canyon National Park provide a better understanding of the health of this population.

This demographic study of Goodding willows in the Grand Canyon also seeks to detect broad scale shifts in Goodding willow geographic distribution from the upper to lower Grand Canyon region. Our study includes Goodding willows samples from the entire stretch of the Colorado River through Grand Canyon National Park. The three main regions containing populations of Goodding willows are: (1) Lees Ferry (river mile 0), downstream from the Glen Canyon dam; (2) Cardenas Creek (around river mile 71), close to the Little Colorado River confluence; and (3) upstream from Pearce Ferry (river mile 267-274), in the Upper Lake Mead region created by the Hoover Dam. A fourth site at Diamond Creek, a side canyon with Goodding willows upstream from the Colorado River (river mile 226), serves as a control site not influenced by river flow regulations. The survey also includes all found scattered Goodding willows accessible by boat along the Colorado River in Grand Canyon National Park. At each field site, permanent plots are noted with numbered tags on cored willows and plot boundaries are marked on aerial photographs. These permanent plots help in long-term research on Goodding willow regeneration, establishment, and mortality along the Colorado River in Grand Canyon National Park.

Ecological information in this study includes taking tree cores from the base of Goodding willows and measuring tree diameters at ground height. Tree condition data consists of stem counts, number of dead stems, and beaver or flood damage. Annual rings in mounted tree cores help to determine tree age, detect of periods of growth suppression and release, and determine of initial growth. In addition to visually cross-dating the tree core samples, tree-ring computer programs enable us to determine ring widths using an incremental measuring machine and microscope. Tree-ring statistical programs then help us to detect dating problems, such as false or missing rings. Master chronologies for each of the main willow areas allow the examination of climatic, disturbance, and river flow influences on tree growth.

Comparing the three river sites, the results show the Lees Ferry upstream site near the Glen Canyon Dam contains the oldest Goodding willows dating back to the 1930s, with few young trees. In terms of the health of the population at Lees Ferry, half of the Goodding willows have dead stems and 20% exhibit beaver damage. Further down the river at the Cardenas Creek site, a few Goodding willows date back to the 1950s, but the majority of establishment dates to just after 1973, the year of an increase in maximum river flow. The Pearce Ferry downstream site in the Upper Lake Mead region has 60% of the Goodding willows establish following the large 1983 flood, and none have dead stems. Similar to Pearce Ferry, the control side-canyon site of Diamond Creek (off of river mile 226) has Goodding willows dating back only to 1973, with no dead stems found.

When analyzing the diameter of the trees by site, the Lees Ferry area upstream site possess the greatest number of large diameter Goodding willows, up to 87 cm. At Cardenas Creek, one Goodding willow reaches 85 cm, but the majority (78%) are smaller and range from 10-25 cm. In contrast, at the downstream site of Pearce Ferry along Upper Lake Mead, the largest diameter Goodding willow in the sample area measures only 33 cm, with almost all Goodding willows (91%) under 20 cm. Similarly, at the control side canyon site of Diamond Creek almost all Goodding willows are under 20 cm with only one large tree (at 55 cm).

Comparing sites along the Colorado River, results from master chronologies show impacts of the dam construction through suppressions of willow growth. At the upstream area of Lees Ferry, significant growth suppression occurred starting in 1964, a year after dam completion. Goodding willows farther downstream at Cardenas Creek show suppression seven years after dam establishment. Delayed reactions may be due to the mitigating effects of the Little Colorado River that junctions with the Colorado River nine miles upstream from Cardenas Creek. Similarly, the Goodding willows at the Pearce Ferry site in Upper Lake Mead display a marked suppression starting in 1971. This may be a delayed reaction to Glen Canyon dam, or possibly a result of fluctuating levels in Lake Mead.

With regards to increases in willow growth rates, Cardenas Creek's show a significant increase in growth after the 1983 post-dam record flood. These Goodding willows display a clear growth release starting in 1985, likely reflecting the flood in 1983 and above average maximum flows in 1984-86. The control-site of Diamond Creek off the main Colorado River displays no evidence of releases following the 1983 flood, an anticipated result since Colorado River flow fluctuations should not influence these willows in a side canyon. Instead, the Diamond Creek side canyon site shows potential climate-induced growth trends. The periods of suppression from climate include both 1976-77 and 1980-82, while periods of release occur in 1975 and 1978. In contrast, none of the three Colorado River sites display significant suppression during 1976-77 or 1980-82, nor exhibit increased growth in 1975 or 1978.

These results illustrate the impacts of the Glen Canyon dam on the Goodding willow, the largest native riparian tree in the Grand Canyon. These findings stress the importance of flood events like the 1983 record post dam flood for establishing Goodding willows, as well as impacts of regulated river flows suppressing willow growth. For example, at the upstream area of Lees Ferry, significant suppression in Goodding willow growth occurred a year after the Glen Canyon Dam completion. This site has little willow regeneration and the older pre-dam willows are dying. In contrast, the midstream area of Cardenas Creek's master chronology shows a significant increase in Goodding willow growth starting in 1985, reflecting the impact of the 1983 flood and above average maximum flows in 1984-86.

Moreover, our demographic studies quantifies the broad scale geographic shifts in the Goodding willow species distribution from the upper canyon (Lees Ferry) area to lower canyon (Pearce Ferry) region of Upper Lake Mead. Little recruitment and high mortality levels are evident in upstream populations, comprised mainly of decadent individuals. Even though most species have increased their distribution since dam construction, Goodding willow appears not to be recruiting in most of the Grand Canyon National Park. Instead, Goodding willows are establishing at downstream sites near the Hoover Dam in Upper Lake Mead on newly established benches, possibly due to ample exposed benches of sediments. Yet even this population is in danger from large lake level fluctuations and droughts stranding the trees far above the lake.

The negative impact of Glen Canyon Dam on Goodding Willow along the riparian corridor of the Grand Canyon National Park raises several important management issues. The Park Service is bound by its Organic Act to manage its natural resources, including physical and biological processes as well as individual species and plant and animal communities, in a manner that leaves them unimpaired for future generations. Activities that impair a park's natural resources are to be avoided. The primary harm to Goodding Willow, however, comes from operation of the Glen Canyon, which falls outside the jurisdiction of the Park Service.

Conclusion

For the Park Service, managing national parks means managing multiple goals in natural environments that are only partially understood and in political environments that are only partially controlled. The goal of preserving native biodiversity in the Grand Canyon has legal roots in the Park Service's Organic Act, in the ESA, and in the Grand Canyon Protection Act, and is consistent with the Park Service's preservationist mission to safeguard the resources in its care for posterity's sake. But the point of preserving these resources is largely an anthropocentric one; they are to be preserved for our enjoyment and for the enjoyment of generations to come. Thus the interests of recreation legitimately command the attention of planners and managers of our Parks. In determining an appropriate balance between using and preserving these resources, the National Park Service is bounded not just by inherent conflicts between preserving and using a resource, but by the limits of its truncated jurisdiction.

Grand Canyon National Park is legally obligated and organizationally committed to protecting the native species along the riparian corridor of the Colorado River. Much attention has been paid by interest groups and the media to the struggle of native fish like the Colorado pikeminnow, the razorback sucker and the humpback chub. Our research on the Goodding willow suggests that these are not the only species harmed by the ecological changes wrought by the Glen Canyon Dam; to the extent that the conflict over dam management is framed as a 'fish versus power' issue, an incomplete picture of what is at stake is being portrayed.

The ability of the National Park Service to address the root cause of the harm to native species within Grand Canyon National Park is limited by the fact that that root cause is a dam that generates considerable commercial benefits that in turn generate considerable political support for dam operations that maximize those benefits. Moreover, the dam is managed by an agency other than the Park Service, with different, indeed conflicting, legal imperatives, and a different organizational mission. The trade-offs involved reflect both explicit and implicit political choices made in the past, and will inevitably require more political choices in the future. The best way to ensure sound political choices is to understand these trade-offs and education the public about them.

Beyond the conflict over how the Glen Canyon Dam should be managed is the emerging perspective that it shouldn't exist at all. Calls to de-commission the Glen Canyon Dam and return the Colorado River to its natural state remain on the margins of the public debate, but history shows marginal voices can become mainstream voices with time. But for those managing the Grand Canyon National Park, an important constituency has emerged in the wake of the Glen Canyon Dam's construction in the form of the whitewater rafting industry which helps the Park Service fulfill its mission to facilitate the public's enjoyment of one of the earth's grandest places. The Colorado River was rafted before the Glen Canyon Dam was built, and it would continue be rafted if it were to be de-commissioned. But certainly dramatic changes would be required of the industry if the Colorado River was restored to a wild condition, and the interests of the industry and those they serve would rightly need to be accounted for.

But until removing the Glen Canyon Dam becomes a more realistic political option, the range of viable management options involves releasing water in ways that restore beaches and sandbars that natural floods used to create at a cost of lost electricity generation, to running the Glen Canyon Dam in ways that maximize its power potential at the cost of scouring the river corridor of its sediment and back-water and near-shore habitat. Also important in this management mix are the Bureau of Reclamation's efforts to divert warmer water from Lake Powell's surface to raise the temperature of the downstream river. Here the interests of biodiversity and recreation may be mutually served, in that humpback chub and Goodding willow as well as river running campers benefit from the beaches and the back-water habitat created by well-planned simulated flooding. Warmer water would be of benefit to native fish and capsized rafter alike.

But managing the Glen Canyon Dam for these purposes will require a more explicit decision than Congress was willing to make in 1992 in the Grand Canyon Protection Act. Are more natural conditions that benefit humpback chub, razorback suckers and Gooding willow worth millions of dollars in lost electricity? This is the choice that is emerging. Direct beneficiaries of the power industry may be unlikely to think so. But the Grand Canyon is an American icon, and a World Heritage Site. Thus the appropriate constituency with a stake in these management issues expands well beyond the region. It is now over 60 years since Aldo Leopold argued that in order to fully respect nonhuman life and the systems on which it depends we must begin to more comprehensively conceptualize our sense of community to encompass the biotic community in both ecological and ethical senses (Leopold, 1948). Are we capable of that? The idea of the National Park system suggests that, in a limited sense at least, we are. The sentiments expressed by Teddy Roosevelt that the Grand Canyon is a unique treasure belonging to all Americans now and to come suggest that if we can conceptualize land and its ecological components as part of our community entitled to ethical consideration, the Grand Canyon is such a place.

Acknowledgements

For help collecting Goodding willow data in the field, we thank Gwen Waring, Kristy Stefan, Melissa Aldrich, Barbara Kent, and Larry Stevens. We also express appreciation to Glen Canyon Environmental Studies and the U.S. Geological Survey (Denver Branch) for river logistics and support. For laboratory assistance, we thank Melissa Aldrich and Barbara Kent. Funding provided through the National Biological Service/National Park Service grant (contract number CA 8045-8-0002).

References

Adler, R.W. 2007. *Restoring Colorado River Ecosystems, A Troubled Sense of Immensity*. Boulder, CO: Island Press.

Bureau of Reclamation, 2009. Managing Water in the West, Upper Colorado Region. Accessed on at: http://www.usbr.gov/uc/water/crsp/cs/gcd.html

Carothers, S.W. & Aitchison, S.W. (Editors) 1976. An ecological survey of the riparian zone of the Colorado River between Lees Ferry and the Grand Wash Cliffs. *Grand Canyon National Park Colorado River Technical Report* **10**. Grand Canyon, Arizona.

Carothers, S.W. & Dolan, R. 1982. Dam changes of the Colorado River. *Natural History Magazine* **91**: 74-83.

Clarke, J.N. & McCool, D.C. 1985. *Staking out the Terrain: Power Differentials Between National Resource Management Agencies*. Albany, NY: SUNY Press.

Clover, E.U. & Jotter, L. 1944. Floristic studies in the canyon of the Colorado and tributaries. *American Midland Naturalist* **32**: 591-642.

Freemuth, J.C. 1991. *Islands Under Siege: National Parks and the Politics of External Threats*. Lawrence, KS: University of Kansas Press.

Ghiglieri, M.P. & Meyers, T.M. 2001. *Over the Edge: Death in Grand Canyon*. Flagstaff, AZ: Puma Press.

Hayes, S.P. 1999. *Conservation and the Gospel of Efficiency: The Progressive Conservation Movement*, 1890-1920. Pittsburgh, PA: University of Pittsburgh Press.

Hayes, S.P. 2000. *A History of Environmental Politics since 1945. Pittsburgh, PA*: University of Pittsburgh Press.

Hjerpe, E.E. & Kim, Y.S. 2007. Regional economic impacts of Grand Canyon river runners. *Journal of Environmental Management* **85**: 137-149.

Howard, A. & Dolan, R. 1981. Geomorphology of the Colorado River in the Grand Canyon. *Journal of Geology* **89**: 269-298.

Leopold, A. 1948. *A Sand County Almanac, and Sketches Here and There*. New York, NY: Oxford University Press.

Mast, J.N. 1997. *Changes in willow populations since Glen Canyon Dam,* Grand Canyon National Park. Pacifica Fall Issue: 1-3.

Mast, J.N. & Waring, G.L. 1996. Chapter 12: *Historical vegetation patterns along the Colorado River in Grand Canyon: changes in Goodding Willow detected with dendrochronology*. Proceedings of the Third Biennial Conference of Research on the Colorado Plateau, pp 115-127.

Morehouse, B.J. 1996. *A Placed Called Grand Canyon: Contested Geographies.* Tucson, AZ: The University of Arizona Press.

Nash, R. 1982. Wilderness and American Mind. New Haven, CT: Yale University Press.

National Park Service, 2006. Colorado River Management Plan, Record of Decision. Accessed online at: http://www.nps.gov/archive/grca/crmp/

National Park Service, 2009. Grand Canyon Annual Visitation 1915-Present. Accessed online at: http://www.nps.gov/grca/parkmgmt/upload/Visitor-Stat-Annual.pdf

National Park Service Public Use Statistics Office, 2009. Accessed online at: http://www.nature.nps.gov/stats/park.cfm

National Research Council. 1996. *River Resource Management in the Grand Canyon.* Washington D.C.: National Academy Press.

Parsons, D.J. 2004. Supporting Ecological Research in the U.S. National Parks: Challenges and Opportunities. *Ecological Applications* **14**: 5-13.

Powell, J.L. 2008. Dead Pool: Lake Powell, *Global Warming, and the Future of Water in the West.* Berkeley, CA: University of California Press.

Purcherelli, M.J. 1986. *Evaluation of riparian vegetation trends in the Grand Canyon using multitemporal remote sensing techniques.* ASPRS Technical Paper. ASPRS-ACSM Fall Convention, pp. 172-181.

Reichenbacher, F.W. 1984. Ecology and evolution of southwestern riparian plant communities. *Desert Plants* **6**: 15-22.

Roosevelt, T.R., 1903. Speech delivered at Grand Canyon, Arizona. published in *A Compilation of the Messages and Speeches of Theodore Roosevelt,* 1901-1905. Lewis, A.H. Editor. 1906. Washington D.C.: Bureau of National Literature and Art.

Runte, A. 1987. *National Parks: The American Experience.* Lincoln, NE: University of Nebraska Press.

Slaughter J.E., Makinster A.S., Speas D.W. & Persons W.R., 2003. Status, trends and long-term monitoring of Lee's Ferry Tailwater Fishery, 1991-2002. Arizona Game and Fish Department, Research Branch Report. Accessed online at: http://www.gcmrc.gov/news_info/outreach/symposiums/2003/cd/pdfs/slaughter_symp03.pdf

Stevens, L. 1983. *The Colorado River in Grand Canyon*: A Guide (4th ed.) Flagstaff, Arizona: Red Lake Books.

Stevens, L.E., & Waring, G.L. 1988. *Effects of post-dam flooding on riparian substrates, vegetation and invertebrate populations in the Colorado River corridor in Grand Canyon,* Arizona. Glen Canyon Environmental Studies. Executive Summaries of Technical Reports. NTIS No. PB88-183488/AS.

Stevens, L.E., Ayers, T.J., Bennett, J.B., Christensen, K., Kearsley, M.J.C., Meretsky, V. J., Phillips, A.M., Parnell, R.A. Spence, J., Sogge, M. K., Springer, A.E., & Wegner, D.L., 2001. Planned Flooding and Colorado River Riparian Trade-Offs Downstream from Glen Canyon Dam, Arizona. *Ecological Applications.* **11**: 701-710.

Stoddard, J.L., 1905. "John L. Stoddard's Lectures Volume 10, 'Grand Cañon of the Colorado River'" from The Project Gutenburg eBook. Accessed online at:http://www.gutenberg.org/files/15526/15526-h/15526-h.htm

Thomas, H.E., Gould, H.R. & Langbein, W.B. 1960. Life of the reservoir. In W.O. Smith et al. Comprehensive survey of sedimentation in Lake Mead, 1948-49. U.S. *Geological*

Survey Professional Paper 295-T. Washington D.C.: U.S. Government Printing Office, pp. 231-248

Turner, R.M., & Karpiscak, M.M. 1980. Recent vegetation changes along the Colorado River between Glen Canyon Dam and Lake Mead, AZ. U.S. *Geological Survey Professional Paper* **1132**. Washington D.C.: U.S. Government Printing Office.

United States Geological Survey, 2008. Science Activities Associated with Proposed 2008 High Flow Experiment at Glen Canyon Dam. Department of the Interior Fact Sheet 2008-3011. Accessed online at: http://pubs.usgs.gov/fs/2008/3011/fs2008-3011.pdf

Wilson, Janet 2008. *Plan to 'flush' Grand Canyon stirs concerns*. Los Angeles Times, March 4, 2008.

In: National Parks: Biodiversity, Conservation and Tourism ISBN: 978-1-60741-465-0
Editors: A. O'Reilly and D. Murphy, pp. 43-60 © 2010 Nova Science Publishers, Inc.

Chapter 3

DETERMINATION OF OPTIMUM MANAGEMENT STRATEGY FOR KÜRE MOUNTAINS NATIONAL PARK IN TURKEY

İsmet Daşdemir[a] and Ersin Güngör[b]
University of Bartın, Faculty of Forestry, Bartın, Turkey

Abstract

There are 39 national parks in Turkey and totally 877.771 hectares area is used for this aim. One of them is Küre Mountains National Park (KMNP), which has 37.000 hectares area and was decelerated in the year 2000. It is the first national park which its boundaries determined by contribution and participation of different society groups in Turkey. In this study, it is aimed to determine optimum management strategy for KMNP by turning out preferences of different society groups (local villagers, public institution and nongovernmental organization representatives, potential tourists) and providing their participation to the management process effectively. To realize this aim, it has been studied to develop a management model taking into consideration social and economic values as well as ecological values. It is also investigated a management model such as *to manage conservation areas by participation principle* and *to use income from conservation areas for developing local villagers*.

In the study, the present situation of the national park is turned out by SWOT analysis, and some alternative management strategies (scenarios) for the national park are developed based on the findings in this stage. The four attributes such as managing type, entrance fee, sharing income and administrative structure which each one has three sub-levels are taken into consideration while developing alternative management scenarios. The nine orthogonal alternative management strategies based on these attributes and their sub-levels are submitted to 462 interviewees selected by layer-random sampling method. The interviewees arranged these strategies according to their preferences. The preference results are evaluated by conjoint analysis, and thus the optimum management strategy is determined for the national park as a managing system with *the balance conservation use, taking US$ 10 entrance fee from visitors, sharing the 70% of income to the national park management and its 30% to local villagers, and administrating with the cooperation of state, nongovernmental*

---

[a] E-mail address: isdasdemir@hotmail.com (İ. Daşdemir); Tel.: +90 378 2277422; Fax: +90 378 2277421. (Corresponding author).
[b] E-mail address: ersingngr@yahoo.com (E. Güngör).

organizations and local villagers. Taking into consideration the optimum management strategy in the management plan of KMNP, which is though preparation, will help the plan to be applicable and dynamic structure, prevent conflicts among local villagers and the national park management, and thus, to contribute to sustainable management of the national park.

Keywords: Küre Mountains National Park, national park management strategy, participation, SWOT analysis, conjoint analysis, Turkey.

1. Introduction

National parks concept together with Yellowstone region had been announced at 1872 in the United States of America (USA) for the first time in the world. After 1960, it is applied intensively and each country has developed its own national park system depending on its natural resources' features, demands, social, economic and cultural structure. National park studies in Turkey had started in 1956 by the 3rd and 25th items of Forest Act (numbered 6831) and they had reached a legal basis in 1983 by National Parks Act (numbered 2873).

Well-done management plans (master plan or long-term management plan) are need for ensuring sustainability, improvement and protection of natural, cultural and recreational resources values of protected areas including national parks and others. To be successful these plans, they should be prepared taking into consideration environmental, social and economical variables. For this, it is need firstly disclosing present situation and determining management aims of national park and their priorities. Later alternatives management strategies (scenarios) should be proposed taking into consideration opinions and contributions of different society groups or interest groups and the best one among them should be selected according to multicriteria and applied to reach the aims. Majority of national parks having main part (82%) in protected area in Turkey do not have management plans yet. The planning mentality mentioned above does not exist in preparation and application stage of management plans in even if some national parks having management plans, and also it is know which is not allowed participation principle to happen in these plans. This causes that plans do not exact work and problems are lived between national park management and interest groups. To provide effective management in protected area, systematical evaluation studies based on participation and scientific bases should be made, a management structure should be constituted for sustainable administration, and interest groups should be effective participated in making decision, planning, applying and control processes. Taking into consideration public benefit, participator and sharer mentality and management models based on social and economic values as well as ecological values are appropriated.

As regards date of 2008, there are 39 national parks in Turkey and totally 877.771 ha area is used for this aim. One of them is Küre Mountains National Park (KMNP), which has 37.000 ha area and was decelerated as a national park in the year 2000. Participation principle was taken basis while decelerating of KMNP having natural and cultural richness and determining its boundaries. However, its management plan could not be prepared as synchronization with its announcement. Also, necessary importance was not given for participation principle in preparation stage of draft management plan. Therefore, alternative management strategies which will provide interest groups to be effective participated in decision making, planning, applying and control processes should be developed as analytic for sustainable management of KMNP, these should be evaluated by multicriteria decision making techniques and the best one (optimum strategy) among them should be selected and this strategy should be taken basis in management plan. Although there are many analytic and multidimensional studies done in this subject in the world (Teeter and Dyer, 1986; Hyberg, 1987; Stevens et al., 2000 etc.), this kind studies in Turkey are few. Even if some studies done in preparation of management plans of protected areas in Turkey (Kuvan, 1997; Menteş, 2001; Demir, 2001; Kalem, 2001) are generally deprived of analytic evaluations and using multicriteria decision making techniques, and they are based on subjective assessments.

There are many multicriteria decision making techniques in the literature. Conjoint analysis is the most important one of them, which is increasing practicability as parallel to improving computer technology and giving an opportunity easier measuring preferences of society. Today in the developed countries, conjoint analysis is commonly used especially in many forestry studies such as to compare forest planning and management strategies (Teeter and Dyer, 1986; Hyberg, 1987; F. C. Zinkhan and G. M. Zinkhan, 1994; Stevens et al., 2000) to determine fundamental principles of multiple use (Zinkhan and Holmes, 1997; Sayadi et al., 2000), to set priorities of nature tourism or ecotourism activities (Morimoto, 1999; Suh and Gartner, 2004), to rank preferences of forest products taken environmental certificate (Bigsby and Ozanne, 2002), to assign protection value of natural resources and to fix damage in these resources (Matnews et al., 1995; Holmes et al., 1996; Holmes et al., 1998; Kuriyama, 1998), to estimate value of non-market forest goods and services (Mackenzie, 1990, 1993; Gan and Luzar, 1993; Adamowicz et al., 1994; Roe et al., 1996). Conjoint analysis is basically used to determine optimum management strategies for KNMP in our study, too. Thus, conjoint analysis is first time used by this study in forestry area in Turkey.

Aim of the study is to determine optimum management strategy for KMNP by a participation approach. For this aim, it has been studied to develop a management model taking into consideration social and economic values as well as ecological values. It is also investigated some management scenarios such as *to manage conservation areas by participation principle* and *to use income from conservation areas for developing local villagers*. Thus, by developing alternative management strategies taking into consideration many attributes, preferences of interest groups related to management strategies are appraised. At the end of evaluating data obtained by interview method with conjoint analysis, firstly the most important attributes and their sub-levels affecting preferences of each interest group are turned out, later the attributes and their sub-levels (part-worth utility, importance value) which are important or most preferred for all interest groups (general) are disclosed. Based on these findings, management strategies preferred by each interest group and the optimum management strategy for KMNP appropriated by all of interest groups (all interviewees) are determined, and its validity from different sides is discussed and interpreted.

2. Material and Method

2.1. Study Area

The study is carried out in KMNP having the oldest forests of Europe, magnificent endemic wildlife, biodiversity, karstic geological structure, canyons, caves and waterfalls rarely observed in the world, and a rich flora, fauna, authentic and folkloric cultural richness

(Figure 1). Hence, World Wildlife Fund (WWF) has listed the area as one of a hundred forest hot spots in Europe deserving priority conservation.

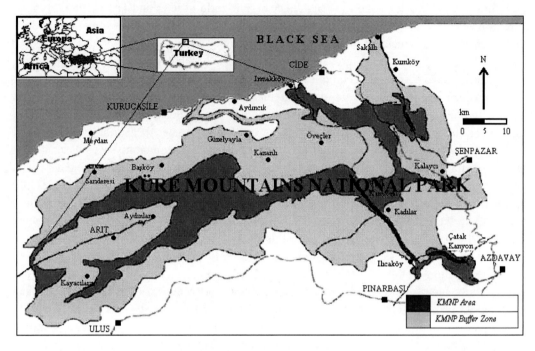

Figure 1. Location of the study area.

KMNP lies between Kastamonu and Bartın, in the Western Black Sea Region of Turkey. Total area (black area in Figure 1) of the national park is 37.000 ha, its 17.000 ha is in the boundaries of Bartın province and its 20.000 ha is in the boundaries of Kastamonu province. Therefore, it is called as name of Kastamonu-Bartın Küre National Park in some literature, too. Also, the 80.000 ha area around of the national park is divided as *buffer zone*. Although there is not any settlement in area of the national park, there are many settlements in buffer zone. Because there is no definition of buffer zone is in National Park Act with numbered 2873, it could not be reached a legal basis. Therefore, only 37.000 ha part of the area is divided into three different zones inside of its own as (absolute conservation zone, recreational usage zone, rehabilitation zone) and it is decelerated as a national park. Draft management plan of the national park has not been accepted yet. It is jointly managed by Kastamonu and Bartın Environment and Forest Directorates as administrative.

2.2. Data and Design of the Study

In the study, firs of all, the relevant literatures (Akesen, 1978, 1998; Kuvan, 1997; Kalem, 2001; Menteş, 2001; DKMPGM, 1999; Demir, 2001; Zal, 2002; Daşdemir and Akça, 2002; Bartın Environment and Forest Directorate, 2003; WWF-Turkey, 2003; Karabıyık and Çetinkaya, 2003; Gezi Travel, 2004; Karabıyık, 2004; Daşdemir and Güngör, 2005; National Geopraphic, 2005; PANPARKS, 2005; EUROPARK, 2005, UNDP, 2005; FAO, 2005) were inspected. By taking into consideration findings obtained in this stage too, the four attributes such as *managing type, entrance fee, sharing income* and *administrative structure* thought to be important for managing of the national park (NP) and each one has three sub-level, were decided (Table 1).

Table 1. Decided attributes and their sub-levels.

Attribute	Attribute sub-level	
	Number	Description
(1) *Managing type*	1–1	Conservation
	1–2	Conservation + Use
	1–3	Use
(2) *Entrance fee*	2–1	US$ 3.00
	2–2	US$ 5.00
	2–3	US$ 10.00
(3) *Sharing income*	3–1	100 % State
	3–2	25% State + 50% National Park + 25% Local Villagers
	3–3	70% National Park + 30% Local Villagers
(4) *Administrative structure*	4–1	Only State
	4–2	State + Local Villagers + NGO
	4–3	Only Local Villagers

By taking these attributes and their sub-levels into consideration, according to the full-profile approach (design), it is possible to be developed totally 81 (=3^4) combinations (management strategy or scenario) (Green and Sirinavasan, 1978; Malhotra, 1996). However, in this case, because that interviewees give a score to 81 combinations or range them is difficult and takes a lot of time, according to the orthogonal array solution which is special sub-class of full-profile design (Hair et al., 1995; Smith, 1999), the nine combinations or management strategies which correlations are zero among them were developed by based on the attributes and their sub-levels in Table 1.

Firstly, the cards showing main characteristics of each one of developed management strategies were prepared. Later these cards with together a questionnaire measuring some socio-economic features of interviewees and explaining meanings of the attributes and their sub-levels were submitted to interviewees by interview method. Ranking the management strategies from the best to the worst and rating them by interviewees was provided. These rating were obtained using a nine-point Likert scale (1= not preferred, 9= greatly preferred) Questionnaire and interview studies were carried out in Kastamonu and Bartın provinces in 2005.

Because structure of target social group is heterogeneous in the study, firstly it was dived to four sub-layers or interest group (local villagers (LV), public institution and nongovernmental organization (NGO) representatives, potential tourists) and some interviewees were selected from each interest group by random sampling method. Thus, the *layer-simple random sampling method* was used (Kalıpsız, 1987; 1994). To interview with at least 100 people from each interest group was aimed to get reliable results and to make possibility using statistical analyses. For this aim, it was interviewed with totally 462 interviewees who consist of 145 local villagers, 112 public institution representatives, 101 NGO representatives, and 104 potential tourists. 145 local villagers are equal to 1% of total 14.479 people living in 68 the villages (31 of them in Bartın, 37 of them in Kastamonu) in buffer zone of KMNP (SSI, 2000). Similarly, 112 public institution representatives are equal to 10% of 1120 public officials working in Bartın and Kastamonu provinces. In the same way, 101 NGO representatives are equal to 10% of 1100 NGO members in Bartın and Kastamonu provinces and 104 potential tourists are equal to 0.1% of 181.032 visitors coming for the year 2003 to the region (Bartın Governorship, 2005; Kastamonu Governorship, 2004).

2.3. Evaluation of Data

In the study, the present situation of the national park is turned out by SWOT (Strength, Weakness, Opportunity and Threat) analysis based on the results from investigating of many literatures and land surveys. To evaluate data obtained by interview method and to choose the best one among nine alternative management strategies, conjoint analysis which is a multivariable optimization technique was used. Generally, the basic conjoint analysis model used to measure preferences of consumers towards any product or service may be represented by the following formula (Malhotra, 1996; Daşdemir, 2005a):

$$U(X) = \sum_{i=1}^{m} \sum_{j=1}^{k_j} a_{ij} X_{ij}$$

Where;
U(X) : Overall utility of an alternative,
a_{ij} : The part-worth utility associated with the jth sub-level (j=1,2,...., k_i) the of ith attribute (i=1,2,....,m),
X_{ij} : Dummy variable taking 1 value for the jth sub-level of the ith attribute and 0 value for others,
k_i : Number of sub-levels of the ith attribute (j=1,2,..., k_i),
m : Number of attribute (i=1,2,....,m) are shown.

Preference (overall utility) function is generally handled as a dependent variable and effects of many independent variables on it are investigated in conjoint analysis using qualitative and quantitative data and including some statistical analysis such as correlation and regression too. Thus, effect of each independent variable on preference structure of consumer is determined (Tatlıdil, 1995). In this study, conjoint analysis program was written in the File-New-Syntax menu of SPSS (Statistical Package for Social Science) 9.0 and solutions were made by this program.

3. Results and Discussion

3.1. SWOT Analysis for KMNP

At the end of SWOT analysis, strengths and weakness of KMNP, and opportunities and threats caused by external environment conditions are determined as those in Table 2.

Table 2. Results of SWOT analysis for KMNP.

SWOT Analysis	
STRENGTHS	**WEAKNESSES**
Existence of the Draft Master Plan to be a base for Management Plan, Its boundaries were determined by contribution and participation of interest groups, Existence of authentic and folkloric cultural richness, Existence of natural, old, virginal and mixed forests, biodiversity and magnificent endemic wildlife, Existence of karstic geological structure, canyons, caves and waterfalls rarely observed in the world, Richness of flora and fauna.	Nonexistence of Management Plan, Untidiness and wideness of the national park area, Insufficient equipment and staff, Illegal wood utilizing from its forests, Insufficiency of scientific researches in the area, Not taking of entrance fee to the national park, Not translating of the national park incomes to local villagers because of legal barriers.
OPPORTUNITIES	**THREATS**
Existence of possibility to reach to sensible and conscious visitors mass towards nature, Existence of possibility to be member international certificate institutions as PAN Parks and EUROPARK, Being of one of nine hot spots of Turkey, which are announced by WWF, Existence of two forest faculties around of the national park, Existence of volunteering and willingness for institutional cooperation, Existence of international institutions supporting to its management and finance such as UNDP, FAO, WWF, GEF and JICA*.	Existence of some problems caused by certain acts Insufficiency of nature conservation politics and increase of press towards protected areas Insufficiencies of finance, Insufficiently understanding of importance of protected areas, Existence of uncontrolled tourism activities towards the area, Unsolved cadastral problems in the buffer zone, Not well analyzing of constructing dams, roads and excretory systems of urban hard waste, Uncontrolled and unconscious gathering of some endemic plant species in the area, Existence of excessive rural poverty in the region.

* UNDP: **United Nations Development Program**; FAO: Food and Agriculture Organization; WWF: World Wildlife Fund; GEF: Global Environment Facility, JICA: Japon International Cooperation Agency.

3.2. Some Socio-economic Features of Interviewees

Some socio-economic features of the 462 interviewees interviewed are given in Table 3. According to these, majority of interviewees is male, married and in group of middle age. In case of an evaluation is made according to their education level, a great majority (80%) of public institution representatives graduated from university while 70% of local villagers graduated from primary school. As regards income level, monthly income level of most (65%) of interviewees is between US$ 200 and 800. In spite of this, it is understand that 72% of local villagers has a monthly income lower than US$ 400, 87% of public institution representatives, 48% of NGO representatives and 62% of potential tourists have a monthly income more than US$ 600. Finally, in an evaluation made with regard to occupations; 51%

of local villagers is farmer and retired, 54% of potential tourists is public staff and self-employed person (Table 3).

Table 3. Some socio-economic features of interviewees.

Feature	Level	Local Villagers		Public Institution Representatives		NGO Representatives		Potential Tourists		Total	
		Number	%	Number	%	Number	%	Number	%	Number	%
Sex	Male	110	76	91	81	66	65	49	47	316	68
	Female	35	24	21	19	35	35	55	53	146	32
Age	18-25	15	10	4	4	7	7	20	20	46	11
	26-35	11	8	38	34	19	19	18	17	86	19
	36-45	25	17	41	37	23	23	21	20	110	24
	46-558	43	30	20	18	30	30	22	21	115	25
	56-65	26	18	9	7	13	13	14	13	62	13
	66 +	25	17	0	0	9	8	9	9	43	8
Marital Statue	Married	118	81	93	83	77	76	53	51	341	74
	Single	27	19	19	17	24	24	51	49	121	26
Education Level	Illiterate	18	12	0	0	0	0	2	0	18	3
	Primary school	101	70	1	1	29	27	10	2	133	29
	Secondary school	16	11	1	1	14	15	28	10	41	10
	High school	10	7	20	18	23	23	52	26	81	19
	University	0	0	90	80	35	35	12	62	189	39
Monthly Income	≤ US$ 199	53	37	0	0	11	11	5	5	69	15
	US$ 200-399	53	37	0	0	13	13	11	11	77	17
	US$ 400-599	31	21	14	13	29	28	23	22	97	20
	US$ 600-799	5	3	39	35	15	15	27	26	86	18
	US$ 800-1199	3	2	34	29	18	18	16	15	71	16
	≥ US$ 1200	0	0	25	23	15	15	22	21	62	14
Occupation	Staff of public institution	2	2	108	96	16	16	27	26	152	34
	Personnel of private sector	4	3	0	0	11	11	5	5	20	4
	Self-employed person	6	4	3	3	33	32	31	29	73	15
	Retired	34	23	0	0	13	13	12	12	59	13
	Housewife	17	12	0	0	5	5	9	9	14	7
	Student	6	4	1	1	1	1	16	15	18	5
	Farmer	42	29	0	0	5	5	0	0	5	10
	Unemployed	9	6	0	0	5	5	0	0	5	3
	Other	25	17	0	0	12	12	4	4	16	9

3.3. Management Strategies for KMNP

3.3.1. Developing Alternative Management Strategies

Some alternative management strategies (scenarios) for the national park are developed based on the findings obtained hitherto in this stage. For this aim, the four attributes such as managing type, entrance fee, sharing income and administrative structure thought to be important for managing of the national park and each one has three sub-level in Table 1 were used while developing alternative management scenarios. The nine management strategies (profiles) which correlations are zero among them were developed by based on these attributes and their sub-levels according to the orthogonal array solution (Table 4).

Table 4. The nine alternative management strategies constructed to according to the orthogonal array.

Strategy No	Attributes and their sub-levels			
	Managing type	Entrance fee (US$)	Sharing income	Administrative structure
1	Conservation	10	25% State + 50% NP + 25% LV	Only LV
2	Conservation	5	70% NP + 30% LV	State + LV + NGO
3	Conservation + Use	3	25% State + 50% NP + 25% LV	State + LV + NGO
4	Conservation	3	100% State	Only State
5	Use	10	100% State	State + LV + NGO
6	Use	5	25% State + 50% NP + 25% LV	Only State
7	Conservation + Use	5	100% State	Only LV
8	Conservation + Use	10	70% NP + 30% LV	Only State
9	Use	3	70% NP + 30% LV	Only LV

3.3.2. Evaluation of Preferences by Conjoint Analysis

After preparing the special cards related to the above nine strategies, they were submitted to interviewees by interview method, and the interviewees gave the preference ratings changing between 1 and 9 to them for evaluating by conjoint analysis. In accordance with conjoint analysis, the worst sub-level of each attribute from the viewpoint of the study aim (*use* for managing type, *US$ 10* for entrance fee, *100% state* for sharing income, and only *local villagers* for administrative structure) was taken as the base (reference) levels, and the eight dummy variables taking the value 0 or 1 were defined for the remainder sub-levels of attributes. Assuming the average of preference ratings of 462 interviewees was as the dependent variable and the attributes sub-levels with dummy variable were as the independent variables, the constructed ordinary least squares (OLS) regression model was solved by conjoint analysis and the results are shown in Table 5 and Figure 2. Thus, the part-worth utilities of the attribute sub-levels and the importance values of the attributes which are important for selecting management strategy are determined as based on preferences of each interest group.

Table 5. Results of conjoint analysis showing part-worth utilities of attribute sub-levels and importance values of attributes.

Attribute	Attribute sub-level	Part-worth utility					Importance value (%)				
		LV	Public	NGO	Tourist	General	LV	Public	NGO	Tourist	General
1. *Managing type*	1–1	-0.22	-0.04	-0.52	-0.46	-0.30					
	1–2	0.23	1.61	1.55	1.50	1.14	17.53	47.79	44.14	44.72	36.87
	1–3	-0.01	-1.57	-1.03	-1.04	-0.84					
2. *Entrance fee*	2–1	-0.61	-0.13	-0.23	-0.05	-0.28					
	2–2	0.02	0.04	-0.01	0.16	0.05	16.38	5.36	8.12	19.81	12.67
	2–3	0.59	0.09	0.22	-0.11	0.23					
3. *Sharing income*	3–1	-1.16	-0.45	-0.95	-0.44	-0.78					
	3–2	0.53	0.20	0.13	0.28	0.31	36.67	11.76	24.30	14.27	22.86
	3–3	0.63	0.25	0.82	0.16	0.47					
4. *Administrative structure*	4–1	-0.79	-0.36	-0.67	-0.16	-0.52					
	4–2	0.69	0.99	0.80	0.67	0.78	29.42	35.09	23.44	21.20	27.60
	4–3	0.10	-0.63	-0.13	-0.51	-0.26					

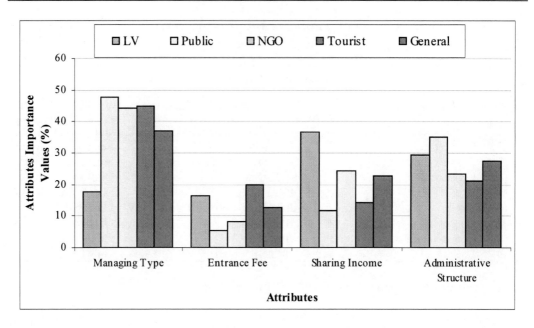

Figure 2. The importance values of attributes from the viewpoint of each interest group and all (general) of interest groups.

3.3.3. Evaluating Attributes and Their Sub-levels According to Each Interest Group

The importance value for each attribute represents the contribution of the attribute to preference for each interest group. According to this, LV, which is one of the four interest groups in the study, arranged the attributes which are important to manage the national park as *sharing income, administrative structure, managing type* and *entrance fee*. Representatives of public institution gave priorities them as *managing type, administrative structure, sharing income* and *entrance fee*. On the other hand, potential tourists arrayed them as *managing type, administrative structure, entrance fee* and *sharing income* while NGO representatives ranged them as *managing type, sharing income, administrative structure* and *entrance fee* (Table 5 and Figure 3). These results shows those local villagers firstly thought getting a share from income obtained by the national park, as for representatives of public institution and NGO and potential tourists did not have such thought and they more desired to be managed as a sustainable of the national park.

The part-worth utility indicates the relative importance or contribution to the overall worth of each sub-level of each attribute from viewpoint of preference of each interest group. According to this, in case of an evaluation will be made in respect to the part-worth utilities of sub-levels of attributes as regards interest groups (Table 5 and Figure 4); it is understood that all of interest groups (general) preferred at most the "conservation + use" sub-level of managing type attribute and at least its "only use or conservation" sub-level. In respect of entrance fee, while LV and representatives of public institution and NGO preferred at most the US$ 10 sub-level of entrance fee attribute and at least its US$ 3 sub-level, potential tourists preferred at most its US$ 5 sub-level and at least its US$ 10 sub-level.

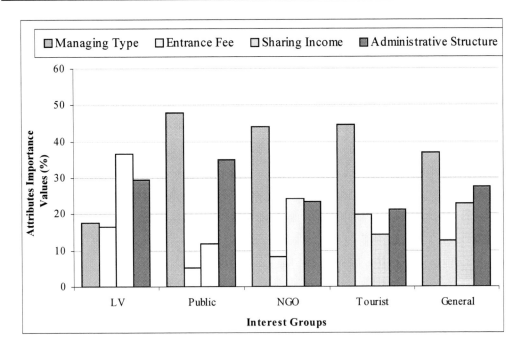

Figure 3. Ranking attributes according to interest groups.

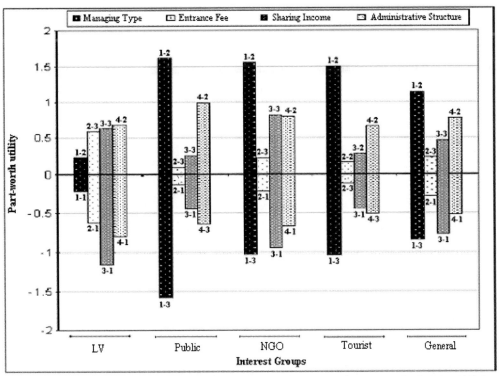

1-1: Conservation, **1-2:** Conservation + Use, **1-3:** Use; **2-1:** US$ 3, **2-2:** US$ 5, **2-3:** US$ 10; **3-1:** 100% State, **3-2:** 25% State + 50% NP + 25% LV, **3-3:** 70% NP + 30% LV; **4-1:** Only State, **4-2:** State + LV + NGO, **4-3:** Only LV

Figure 4. Evaluating attributes' sub-levels in respect to interest groups.

From point of view of sharing income, while LV and representatives of public institution and NGO preferred at most the "70% NP + 30% LV" sub-level of sharing income attribute and at least its "100% state" sub-level, potential tourists preferred at most the "25% state + 50% NP + 25% LV" sub-level and at least its "100% state" sub-level. With respect to administrative structure, while all of interest groups preferred at most to be managed with cooperation of "state + LV + NGO", LV and representatives of NGO preferred at least its "only state" sub-level, and representatives of public institution and potential tourists preferred at least its "Only LV" sub-level.

These results show that whole interest groups agree on the sub-level of managing type, also LV and representatives of public institution and NGO have the same idea on entrance fee, sharing income and administrative structure. As for potential tourists, who basically constitute visitor potential of the national park, are logically against to especially high entrance fee (US$ 10) for their benefit. It was determined that mid-level entrance fee was preferred in some studies made in this subject too (Zinkhan et al., 1997; Holmes et al., 1998; Stevens et al., 2000).

3.3.4. Evaluating Attributes and Their Sub-levels According to All of Interest Groups

In case of an evaluation will be made according to all of interest groups (all interviewees or general); they arranged the attributes which are important to manage the national park as *managing type, administrative structure, sharing income* and *entrance fee* (Figure 3). This ranking is the same with the array made by representatives of public institution. According to this, "managing type" with the 36.87% importance value is the most important attribute in the stage of determining management strategy for the national park. Nevertheless, managing type attribute was preferred at most (47.79%) by representatives of public institution and at least (17.53%) by LV. As for "administrative structure" secondary ranked (27.60%) by all interviewees and this attribute was preferred at most (35.09%) by representatives of public institution and at least (21.20%) by potential tourists. "Sharing income" in third rank with the 22.86% importance value was preferred at most (36.67%) by LV and at least (11.76%) by representatives of public institution. "Entrance fee" in fourth rank was preferred at most (19.81%) by potential tourists and at least (5.36%) by representatives of public institution (Table 5 and Figure 2).

When taking into consideration the attributes' sub-levels and their part-worth utilities with respect to all interviewees (general); it is understood that the "conservation + use" sub-level with 1.14 part-worth utility in "managing type" attribute located in the first rank was preferred at most and its "only conservation" or "only use" sub-levels was preferred at least. This preference ranking is not different among the interest groups. This result implies that the national park should be managed not "only use or conservation" idea but also "conservation + use" idea which combines both of them as balanced, and as suitable for sustainability principle.

The "state + LV + NGO" sub-level with 0.78 part-worth utility in "administrative structure" attribute having a second degree importance to determine management strategy, was mostly preferred by all interviewees. Its "only LV" and "only state" sub-levels was preferred in the second rank and in the last rank respectively. These results imply that the national park should be managed with a cooperation of state, NGO and LV. In fact, it can be said that there is a thought, including common participating to decisions and responsibilities and contributing for more conserving and managing of the national park, in basis of this preference. Participation principle to management is effective in different areas of management in democratic societies as a needed of democracy. One of these areas is management of protected areas (i.e., national park, nature park) too and KMNP will reach a sustainable management because of participation principle.

The "30% NP + 30% LV" sub-level with 0.47 part-worth utility in "sharing income" attribute located in the third rank was firstly preferred by all interviewees. Hence, all interviewees think that transferring the 70% and 30% of the national park's incomes to the management of NP and LV respectively is a rational way in order to manage the national park as the best, to increase quality of services from the park and to develop local villagers as proportional with income. Because all of local villagers living around of the national park are forest villager, they had worked in wood production works in the forest and gained more or less income before decelerating as a national park of the area. Therefore, the interviewees desire local stakeholder to continue getting income after decelerating as a national park of the area too. Subsequently, sharing of incomes obtained from the area of the national park between the national park management and local villagers more preferred rather than the others.

The "US$ 10" sub-level with 0.23 part-worth utility of "entrance fee" attribute located in the fourth rank was preferred at most and its "US$ 3" sub-level with -0.28 part-worth utility at least was desired by all interviewees. While without giving any guide service and the presentation catalog of the national park in the "US$ 3" sub-level choice, both a guide service and the presentation catalog of the national park are given in the "US$ 10" sub-level choice. On the other hand, in the study, the result that potential tourists (or visitor) preferred at most a mid-level entrance fee (US$ 5) resembles results of the research in this subject (Zinkhan et al., 1997; Holmes et al., 1998; Stevens et al., 2000). Main reason of being preferred the "US$ 10" sub-level at most by all interviewees (general) in this study is both to give more quality service to visitor and to transfer a greater share from income of the national park to local villagers having rural poverty and economic difficulties.

3.3.5. Optimum Management Strategy According to Each Interest Group and All of Them

In the study, by taking into consideration the attributes and their sub-levels mostly preferred by each interest group (Table 5), the optimum management strategies for the KMNP, which are appropriated by each interest group (LV, public institution and NGO representatives, potential tourists) and all of them, are determined as follows:

1. *Management strategy appropriated by LV:* It was concluded as a strategy including the managing balance "conservation + use" of the national park, taking "US$ 10 entrance fee", transferring "70% of the national park income to the management of the park and its 30% to LV" and administrating the national park with the cooperation of state, NGO and LV.
2. *Management strategy appropriated by public institution representatives:* This strategy is the same with the strategy appropriated by LV. That is to say, public institution representatives have the same thought with LV for the management of

KMNP. However, public institution representatives give an importance to managing type in the first rank while LV gives an importance to "sharing income" in the first rank.
3. *Management strategy appropriated by NGO representatives:* The strategy fixed by this interest group is the same with the strategies appropriated by LV and NGO representatives. But, the importance degrees of the attributes in the strategy are different as can be seen in Table 5.
4. *Management strategy appropriated by potential tourists:* It was determined as a strategy enclosing the managing balance "conservation + use" of the national park, taking "US$ 5" as entrance fee, transferring "25% of the national park income to state + its 50% to the management of the park + its 30% to LV" and administrating the national park with the cooperation of state, NGO and LV. These results bring up that potential tourists thought different from LV, public institution and NGO representatives for entrance fee.
5. *Management strategy appropriated by all of interest groups:* This is an *optimum* strategy which should be applied in the national park for all interest groups. According to this strategy;

- The national park should be managed with the balance "conservation + use",
- "US$ 10 entrance fee" should be taken from visitors for entrances to the national park,
- 70% of the national park income to the management of the park and its 30% to LV should be transferred,
- The national park should be administrated with the cooperation of state, NGO and LV.

As understood from above explanations, the "conservation + use" sub-level was found important in all of the strategies determined in the study. That this mentality (or result) served sustainable principle was found in some similar researches made in this subject too (Zinkhan et al., 1997; Holmes et al., 1998; Stevens et al., 2000).

4. Conclusion

In this study, which it has handled to determine the optimum management strategy for KMNP by means of a participatory approach, firstly of all, the present situation of the national park is put forth by SWOT analysis. According to the results of SWOT analysis, the most strong sides of KMNP are that it has the oldest forests of Europe, magnificent endemic wildlife, biodiversity, karstic geological structure, canyons, caves and waterfalls rarely observed in the world, and a rich flora, fauna, authentic and folkloric cultural richness. Based on the findings of SWOT analysis too, the nine alternative management strategies which consist of orthogonal array of the four attributes such as managing type, entrance fee, sharing income and administrative structure which each one has three sub-levels, were developed for the national park. They were submitted to 462 interviewees who were selected by layer-simple random sampling method and their majority is male, married and middle aged. Obtained data and their preferences were evaluated by conjoint analysis, and thus the optimum management strategy preferred mostly was determined for KMNP. According this, the optimum management strategy determined by all of interest groups are explained as follows:

> "A managing system with the balance conservation use should be appropriated in the national park, visitor entrances should be limited to the absolute conservation zone, tourism activities (ecotourism etc.) and visitor number (carrying capacity) in the recreational usage zone should be well-planned, "US$ 10 entrance fee" including both a guide service and the presentation catalog should be taken from visitors for entrances to the national park, 70% and 30% of the national park incomes should be transferred for the national park's needs and to local villagers' needs respectively without any income share are transferred to state, the national park should be administrated with the cooperation of state, nongovernmental organizations and local villagers."

As contrary to the today's management mentality, this optimum and general strategy envisages representatives of nongovernmental organizations and local villagers to participate in the management of the national park as well as state. Also, it anticipates an innovation as to be transferred a part of the national park's incomes to local villagers. By means of these innovations, the national park will be more effective managed because conflicts can be prevented, participation in the management can be provided and some sources can be transferred for rural development. On the other hand, operational planning works (Daşdemir, 2005b) and detailed application plans should be made in accordance with framework of this general strategy. For example, some detail issues should be located in application plans an operational planning works such as numbers and locations of temporary constructions such as tent and cabin giving possibility to visitors to make camping without giving any damage to natural cover should be planned; an entrance fee elasticity that does not give permission to overcome its carrying capacity should be applied and tourist guides should be selected people who relatively educated and recognized the region.

Consequently, if the optimum management strategy put forth for the national park is taken into consideration in the preparation stage of the management plan of KMNP, which is though preparation, the plan will be an applicable and dynamic structure; conflicts will be prevented among local villagers and the national park management by means of participation. Thus, a contribution will be provided for protection of the natural, geological, ecological and cultural values of the national park and its sustainable management to be successful.

References

Adomowicz, W., J. Louviere, and M. Williams. 1994. Combining revealed and stated preference methods for valuing environmental amenities. *Journal of Environmental Economics and Management,* **26**: 271-292.

Akesen, A. 1978. *Characteristics and problems of the national parks in Turkey with respect to the outdoor recreation* (Exam.: The Uludağ National Park). İ.U. Publication of the Faculty of Forestry, No: 2484/262, İstanbul, 204 pp.

Akesen, A. 1998. Preliminary evaluation report of protected area project of Valla canyon (TUR7967003). *DKMPGM Report*, Ankara.

Bartın Environment and Forest Directorate, 2003. *Report of environment condition of Bartın city.* Publication of Bartın Governorship, Bartın-Turkey, 143 pp.

Bartın Governorship, 2005. *Report on the conditions of industry, economic and trade of Bartın city in the year* 2004. Publication of Bartın Governorship, Bartın-Turkey.

Bibsby, H.R., and L. K. Ozanne. 2002. *Consumer preference for environmentally certified forest products*: New Zealand and Australia. Lincoln University, Canterbury, New Zealand, 10 pp.

Çemrek, F. 2001. Conjoint analysis using to determine preferences of consumer and an application related to preference of type of credit card. Master Thesis, Osmangazi University, *Natural and Applied Science Institution,* Eskişehir-Turkey, 62 pp.

Daşdemir, İ. 2005a. *Planning and Project Evaluation for Forest Engineering.* Z.K.U. Publication of the Faculty of Forestry, No: 80/16, Bartın-Turkey, 168 pp.

Daşdemir, İ. 2005b. Improving operational planning and management of national parks in Turkey: A case study. *Environmental Management,* **35**(3): 247-257, USA.

Daşdemir, İ., and Y. Akça. 2002. *Some factors affecting the mountain ecosystem of Soğuksu National Park. First National Symposium of Turkey Mountains,* Ministry of Forestry Publication No: 183, Ilgaz -Turkey. pp. 64-70.

Daşdemir, İ., and E. Güngör. 2005. Evaluating national parks in Turkey with regard to tourism certification programs. *Paper of First Environment and Forestry Assembly,* Vol. 4, Antalya, pp.1462-1469.

Demir, C. 2001. Sustainability of tourism and recreational activities in national parks: An application towards the national park in Turkey. Doctorate Thesis, *D.E.U. Social Sciences Institution,* İzmir, 185 pp.

DKMPGM, 1999. Draft Management Plan of Küre Months National Park. Ministry of Forestry. *Publication of National Parks and Hunt-Wildlife General Directorate, UNDP/FAO,* Ankara, 13 pp.

EUROPARKS, 2005. European Federation of Leisure Parks. http://www.europarks.org, 03.01.2005

FAO, 2005. *Food and Agriculture Organization of the United Nations*, http://www.fao.org, 03.01.2005.

Gan, C., and E.J. Luzar. 1993. A conjoint analysis of waterfowl hunting in Louisiana. *Journal of Agricultural and Applied Economics,* **25**: 36-45.

Gezi Travel, 2004. Küre Months National Park. *Journal of Gezi Travel,* Vol. October, İstanbul, pp. 94-107.

Green, P. E., and V. Srinivasan. 1978. Conjoint analysis in consumer research: Issues and outlook. *Journal of Consumer Research,* **5**: 103-123, USA

Hair, J. F., R. E. Anderson, R. L. Tatham, and W.C. Black. 1995. *Multivariate Data Analysis: With Readings*. McMillan Book Company, London, 745 pp.

Holmes, T., C. Zinkhan, K. Alger, and E. Mercer. 1996. Conjoint Analysis of Nature Tourism Values in Bahai, Brazil. *The Forestry Private Enterprise Initiative Working Paper*, No. 57, USA, 19 pp.

Holmes, T., K. Alger, C. Zinkhan, and E. Mercer. 1998. The effect of response time on conjoint analysis estimates of rainforest protection values. *Journal of Forest Economics,* **4**(1): 7-28, USA

Hyberg, B. T. 1987. Multi attribute decision theory and forest management: A discussion and application. *Forest Science,* **33**: 835-845.

Kalem, S. 2001. A new approach in determining tourism potential to be able to protect natural and cultural values and its application in the coast side of Kastamonu city and its near region. Doctorate Thesis, A.U. *Natural and Applied Science Institution*, Ankara, 271 pp.

Kalıpsız, A. 1987. Science and Research. İ.U. *Publication of Natural and Applied Science Institution*, No. 3492/2, İstanbul. 308 pp.

Kalıpsız, A. 1994. *Statistical Methods.* I.U. *Publication of the Faculty of Forestry*, No. 3835/427, İstanbul.

Karabıyık, E. 2004. Marketing research and existing situation of wooden hand products in Kastamonu-Sinop-Bartın and Karabük Cities (in the vicinity of Küre Months National Park). Project of varying and providing sustainability of traditional wood workmanship in Harmangerişi Town of Küre Months, *UNDP-FAO,* Ankara, 35 pp.

Karabıyık, E., and Ö. Çetinkaya. 2003. *Preliminary report of socio-cultural and economic structure. Protection of Biodiversity in Küre Months* National Park and Task I of a Participatory Protected Area Model for Turkey, Ankara, 28 pp.

Kastamonu Governorship, 2004. *Report on the conditions of industry, economic and trade of Kastamonu city in the year 2*003. Publication of Kastamonu Governorship, Kastamonu-Turkey.

Kuriyama, K. 1998. Estimation of the environmental value of recycled wood wastes: A conjoint analysis study. *Forest Economics and Policy Working Paper,* **9801**, USA, 18 pp.

Kuvan, Y. 1997. Issues and solution ways and management of forest recreation resources in Balıkesir Region. Doctorate Thesis, İ.U. *Natural and Applied Science Institution,* İstanbul, 162 pp.

Mackenze, J. 1990. Conjoint analysis of deer hunting, Northeastern. *Journal of Agricultural and Resource Economics,* **19**: 100-107.

Mackenze, J. 1993. A comparison of contingent performance models. *American Journal of Agricultural Economics,* **75**: 593-603.

Malhotra, N. 1996. *Marketing Research: An Applied Orientation*. Prentice-Hall, Inc., USA, 122 pp.

Matnews, K.E., F.R. Johnson, R.W. Dunford, and W.H. Desvousges. 1995. The potential role of conjoint analysis in natural resource damage assessment. *Triangle Economic Research Technical Working Paper,* No. G-9503, USA, pp. 1-18.

Morimoto, S. 1999. *A stated preference study to evaluate the potential for tourism in Luang Prabang,* Laos, Graduate School of International Cooperation Studies, Kobe University, Japan, 20 pp.

Menteş, İ. 2001. Investigating and managing of Ilgaz Month National Park as a protected area. Doctorate Thesis, K.T.U. *Natural and Applied Science Institution,* Trabzon, 270 pp.

National Geographic, 2005. Küre Months, nine hot spots, to protect breakable biosphere of Turkey. *National Geographic-Turkey,* Appendix of February 2005 Number, 20 pp.

PANPARKS, 2005. Protected Area Network of Parks Aims and PANPARKS Criteria, http://www.panparks.org, 03.01.2005.

Roe, B., K. Boyle and M. Teisl. 1996. Using conjoint analysis to derive estimates of compensating variation. *Journal of Environmental Economics and Management,* **31**: 145-159.

Sayadi, S., M.C. Gonzalez, and J. Calatrava. 2000. Ranking versus scale rating in conjoint analysis: Evaluating landscapes in Mountainous Regions in Southeastern Spain, Dpt. *Agricultural Economics CIDA*, Granada, Spain, 19 pp.

Smith, S. 1999. The Concepts of Conjoint Analysis. http://www.marketing.byu.edu/thml/pages/tutorals/conjoint.htm. 01.03.2004.

SSI, 2000. State Statistical Institution 2000 General Population Census, Social and Economic Characteristics of Population. *SSI Publications,* Ankara-Turkey.

Stevens, T.H., R. Belkner, D. Dennis, D. Kittredge, and C. Willis. 2000. Comparison of contingent valuation and conjoint analysis in ecosystem management. *Ecological Economics,* **32**: 63-74, USA.

Suh, Y.K., and W.C. Gartner. 2004. Preferences and trip expenditures- a conjoint analysis of visitors to Seoul, Korea, *Tourism Management,* **25**: 127-137, USA.

Tatlıdil, H. 1995. Conjoint Analysis. Hacettepe University, Department of Statistic, *Lecture Notes,* 25 pp.

Teeter, L. D., and A.L. Dyer. 1986. A multi attribute utility model for incorporating risk in fire management planning. *Forest Science,* **31**: 1032-1048.

UNDP, 2005 United Nations Development Programme. http://www.undp.org.tr, 03.01.2005.

WWF-Turkey, 2003. Presents of Turkey to the World. Publication of WWF-Turkey, Ankara, 8 pp.

Zal, N. 2002. *Following of a project; Küre Months National Park.* First National Symposium of Turkey Mountains, *Ministry of Forestry Publication* No: **183**, Ilgaz -Turkey. pp. 435-441.

Zinkhan, F.C. and G.M. Zinkhan, 1994. An Application of Conjoint Analysis to Capital Budgeting: The Case of Innovative Land Management Systems. *Fin*: **20**, pp. 35-48.

Zinkhan, F.C., and T.P. Holmes. 1997. Conjoint analysis: A preference- based approach for the accounting of multiple benefits in Southern Forest Management. Economics of Forest Protection and Management Research Unit, Research Triangle Park, *SJAF* **21** (4), USA, pp. 180-186.

Chapter 4

THE CONCEPT OF ECOTOURISM DEVELOPMENT WITH FQFD IN THE KINMEN NATIONAL PARK

Tsuen-Ho Hsu[*,a] and Ling-Zhong Lin[≠,b]

[a] National Kaohsiung First University of Science and Technology, Taiwan
[b] Shih Chien University Kaohsiung Campus, Taiwan

Abstract

The paper examines local responses to potential ecotourism development in the Kinmen national park located off the southeastern coast of Fujian Province in Xiamen Bay. Ecotourism management has become an economic issue for tourism industry in conservation and development projects. The traditional approaches of biodiversity conservation concerning areas or national parks protection have been considered ineffective and unethical due to the externally imposed rules and regulations on local people. The establishment of national parks and other forms of protected areas in Taiwan in the past has regarded human activities in such areas as incompatible to the goals of natural resource conservation. Conflicts between the authority of national park and indigenous people were reported during the establishment of national parks in Taiwan, which resulted in the relocation and imposition of constraints on indigenous people.

The development and implementation of ecotourism and integrated conservation development projects have been advocated throughout the study to gain local support. The aim of this study is to propose an approach for the issue of indigenous communities and local government ecotourism management based on the framework of Fuzzy Quality Function Deployment (FQFD), a methodology which has been successfully adopted in analyzing human's attitudes and intention. In particular, this paper addresses the issue of how to deploy the house of quality (HoQ) effectively and efficiently toward key dimensions of ecotourism and tourism guideline of the authority. Fuzzy logic is also adopted to deal with the ill-defined nature of the local residents' attitude preference required in the proposed HoQ. The case of Kinmen national park is presented to demonstrate the implementation of the proposed FQFD

---

[*] E-mail: address: thhsu@ccms.nkfust.edu.tw, Tel: 886-7-6011000 ext. 4217, Address: 1, University Road, Yenchao, Kaohsiung 824, Taiwan, Department of Marketing and Distribution Management, National Kaohsiung First University of Science and Technology, Taiwan.
[≠] E-mail address: ling@mail.kh.usc.edu.tw, Tel: 886-7-6678888 ext. 5712/4251, Address: 200 University Road, Nei-Men Hsiang, Kaohsiung Hsien, 845, Taiwan, Department of Marketing Management, Shih Chien University Kaohsiung Campus, Taiwan.

in ecotourism development. The effective and appropriate management directions for ecotourism development acquired by applying the proposed FQFD, thus, enables the national park authority to achieve an environmental, social, and politico-economic conditions.

Keywords: FQFD, National park, Ecotourism development

1. Introduction

After decades of rapid growth in tourism, China and its neighboring regions in Japan, Korea, and Taiwan have over the past three years experienced sustained increases in the number of tourists visiting their destinations. The research on ecotourism has attracted attention by a number of studies that investigate biodiversity conservation (Brandon, 2001; Hackel, 1999; Nepal & Weber, 1995; Well & Brandon, 1993). Ecotourism and integrated conservation and development projects (ICDPs) are incentive-driven programs applied to link conservation with community development (Spiteri & Nepal, 2006). A common element of various ICDPs is "enhancing biodiversity conservation through approaches which attempt to address the needs, constraints and opportunities of local people" (Well & Brandon, 1993). The ICDPs have been implemented in a variety of forms, including agricultural development, sustainable forestry, wildlife utilization, irrigation and water management, soil enhancement and erosion control, improvement of market access, generating employment opportunities, promoting ecotourism, and providing communities services (Alpert, 1996; Brandon, 2001; Wells & Brandon, 1992).

Similar to the ICDPs, ecotourism's goal is to achieve conservation and community development through the provision of economic and social incentives to local communities (Bookbinder, Dinerstein, Rijal, Cauley & Rajouria, 1998; Chapman, 2003; Wunder, 2000). However, studies show that only few incentive-driven programs have met their goals of balancing biodiversity conservation with the provision of livelihood needs for local communities (Newmark & Hough, 2000; Spiteri & Nepal, 2006; Walpole & Goodwin, 2001). Critiques have argued that generating local support for conservation is challenging because of the highly complex and heterogeneous characteristics of community, and variable interests in issues related to conservation and development (Heinen, 1996). According to Gibson & Marks (1995), one of the factors added the difficulty of incentive-driven programs is that the same incentive or benefit is valued differently by different groups of a community. The situation becomes more complicated when it involves indigenous communities, and their cultural identity and sovereignty (Nepal, 2002; Ryan, 2000).

Brandon & Margoluis (1996), Heinen (1996), and Brandon (2001) suggest that the design of conservation incentive programs should consider not only the environmental conditions, such as the types of protected areas, but also the social and economic contexts of communities so that their values, interests and attitudes can be adequately incorporated in conservation and development plans. However, the measurement of the design of conservation in the ecotourism development has not received much attention. Despite these increased customer-oriented marketing efforts in the tourism industry, relatively little attention has been given to the process of the design of conservation. Although most research programs have focused on measuring tourist perceptions and attitudes of local residents toward tourism development, or traveler perceive risk judgment about tourism development strategy for travel-related products (Allen, Long, Perduce & Kieselbach, 1988; Getz, 1994; Liu & Var, 1986; Long, Perdue & Allen, 1990; Mason & Cheyne, 2000; Smith & Krannich, 1998; Hsu & Lin, 2006). Few have provided guidelines and principles of ecotourism for how to provide the design of conservation to meet the ecotourism development expected by indigenous peoples. Besides, conflicts between local people and national park authorities are often the consequence of externally imposed park regulations. It is suggested that successful national park management will not be achieved without the cooperation and support from local communities and that local communities must be empowered and involved in making important conservation decision (Chi & Wang, 1996). That is, the extant community attitude and community intension were developed primarily as a tool to diagnose an ecotourism development performance and to understand indigenous peoples behavior, but they have not considered actively the incentive-driven programs that can support the guidelines and principles of ecotourism development.

Examination of local attitude toward the key dimensions of ecotourism development helps planners or project managers understand how community members feel about utilizing ecotourism as a means to balance conservation and development. However, measuring community attitude on the design of conservation in the national park is sometimes highly complex, since it incorporates a great variety of uncontrollable and unpredictable factors that affect the local people involved (Heinen, 1996; Newmark & Hough, 2000). Especially, attitude is an individual's favorable or unfavorable disposition toward an object, action, or event of interest (Ajzen & Fishbein, 1980; Fishbein & Manfredo, 1992). It is "a psychological tendency that is expressed by evaluating a particular entity with some degree of favor or disfavor" (Eagly & Chaiken, 1993). Several factors that may complicate the local residents' attitude and preference process, such as incomplete information, additional qualitative criteria and imprecise preferences, are often not taken into account (Ajzen & Fishbein, 1980). Moreover, Hsu and Lin (2006) presented a model that considers the attributes of customer's cognitions by means-end chain analysis and utilizing fuzzy QFD and entropy method helps to structure the amount of information about customer perceived value. The purpose of this paper is to propose for the ecotourism development a systematic, structured approach called Fuzzy Quality Function Deployment (FQFD) adopting an analysis based on fuzzy logic, to provide new extant guidelines and principles of ecotourism development in the national park with a focus on both community members' attitude requirements and the national park authority management requirements. Thus, an understanding of this evaluative response of community members toward ecotourism development helps identify the variation of ecotourism attitude and potential conflicts among community members.

This study is unique in many different ways. First, very few studies have explored community attitude and the planners of the national park toward ecotourism development. Secondly, only limited research has been devoted to exploring community requirements with systematic approach to FQFD prior to tourism or ecotourism development. Although comprehensive studies have been conducted on community attitude toward tourism development, very little research both exists on community attitude and the intention of national park authorities about tourism and conservation. Finally, ecotourism is a relatively new concept in Taiwan where the first nation-wide non-profit ecotourism organization, the Taiwan Ecotourism Association, was established only recently in 2001, to advocated the concept and encourage sustainable practices by the tourism industry and local communities.

A further contribution of this study is to provide insights to the application of internationally defined ecotourism guidelines from a local perspective.

The paper is organized as follows: in Section 2 the FQFD and proposed methodology are discussed. In Section 3, the significance of the tourism industry in Kinmen national park is outlined and issues to the measurement of the design of conservation in the national park is discussed. Section 4 then demonstrates the implementation of the proposed method. Finally, a discussion of the fuzzy-QFD methodology and the conclusion are presented in Section 5.

2. Construction of Fuzzy QFD in Ecotourism Development Delivers

The approach is based on the translation of HoQ (House of Quality) principles from community attitude in dimensions of ecotourism development to guidelines of ecotourism design/management requirements. While the traditional HoQ correlates cutomer' requirements ("whats") with engineering characteristics of new product under development ("hows"), in our approach, community members' requirements in terms of dimensions of ecotourism preference ("whats") are crossed over with viable ecotourism characteristics of the design of conservation management ("hows"). The resulting ecotourism development HoQ is shown in Fig.1.

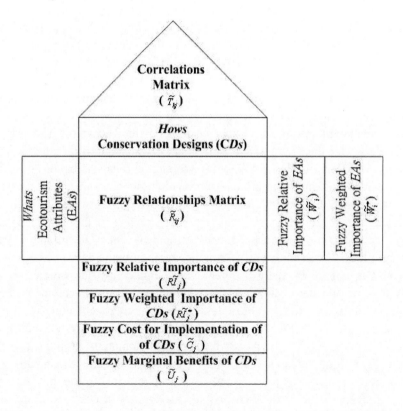

Figure 1. The house of quality for the management of the conservation design.

As can be seen from the Fig. 1, "whats" elements express dimensions of ecotourism attributes EA_i, $i=1, 2, 3,..., n$ affecting the design of conservation management perception. The factors have been extensively described by tourism and environmental management literature. Once dimensions of ecotourism have been asserted, via the design of conservation CD_j, $j=1, 2,..., m$ the authority of the national park could identified and ranked in the conservation field , and improved guidelines and principles of ecotourism performances effectiveness and efficiency. The design of conservation corresponds to "hows" in the proposed ecotourism development for HoQ. Community attitude toward ecotourism requirements are specified on the basis of the tourism development on managerial resource allocation plans in order to satisfy the community members.

The roof of correlations, the weights W_i [$n \times 1$], the relationship matrix R_{ij} [$n \times m$] and the relative importance of CDs vector RI_j [$1 \times m$] complete the HoQ. It is worth stressing that the weight vector, the correlations matrix and the relationships matrix translate local residents' attitude and linguistic preference process given by human beings. Therefore, an effective means to deal with them would appear to be fuzzy logic. Fuzzy logic allows to take into account the different meaning that we may give to the same linguistic expression. This is why the fuzzy approach has been so widely adopted in different research fields. In our approach, four fuzzy elements have been added to the traditional HoQ, namely:

[1] the fuzzy weighted importance of dimensions of ecotourism;
[2] the fuzzy weighted importance of the design of conservation;
[3] (3)the fuzzy cost for the implementation of the design of conservation; and
[4] the fuzzy marginal benefit of the design of conservation.

2.1. Fuzzy Weighted Importance of Dimensions of Ecotourism in Community Members' Attitude

The fuzzy weighted importance $\widetilde{W}_i^*$ of EAs is a [$n \times 1$] vector which expresses the real importance of each EA. The introduction of $\widetilde{W}_i^*$ is required to weight each dimensions of ecotourism considering not only the importance the local residents give it, which is expressed by the fuzzy value $\widetilde{W}_i$, but also the perference delivered by the communities for that dimensions. To correspond with community response, the authority of national park must provide sustainable community development and protected area conservation to the local people on critical dimensions of ecotourism, *that is either those that are perceived as the most important ones or where ecotourism perceived is inferior(delete)*.

The fuzzy weighted importance $\widetilde{W}_i^*$ is computed by assessing the distance d_i between residents' attitudes toward ecotourism devlopment and that which is perceived by residents as superior. Both the attitudes delivered and the target superior attitude value of dimensions of ecotourism could be conducted from fieldwork surveys by asking the local residents directly. Since both attitude values are fuzzy, a distance between fuzzy numbers has to be assessed. To this extent, the hamming procedure is suggested to be adopted (Chien & Tsai, 2000). From a mathematical point of view, given two fuzzy sets $\widetilde{A}$ and $\widetilde{B}$, the Hamming distance

$d(u_{\tilde{A}}(x), u_{\tilde{B}}(x))$ between two fuzzy numbers belonging to $\tilde{A}$ and $\tilde{B}$ respectively, can be computed as

$$d(u_{\tilde{A}}(x), u_{\tilde{B}}(x)) = \int_X |u_{\tilde{A}}(x) - u_{\tilde{B}}(x)| \qquad (1)$$

Where X is the universe of discourse. Due to the calculation method, the resulting Hamming distance is a crisp value.

The result of d_i parameters is calculated from equation (1). Then, the fuzzy weighted importance $\tilde{W}_i^*$ of *EAs* can be derived as equation (2).

$$\tilde{W}_i^* = d_i \otimes \tilde{W}_i \qquad (2)$$

2.2. Fuzzy Relationship Matrix between Dimensions of Ecotourism and the Design of Conservation

The element strives to determine which the design of conservation has the highest impact on local people satisfaction in ecotourism development. It takes into account the fuzzy weighted importance of dimensions of ecotourism, the fuzzy relationships matrix and the correlation matrix.

As already detailed, the position $\tilde{R}_{ij}$ in the relationships matrix expresses the fuzzy relationship between the *j*th *CD* with the *i*th *EA*. Again, a fuzzy linguistic scale may be usefully adopted by the authority of national park to interpret the vagueness and incomplete understanding of the relationships between "hows" and "whats".

The fuzzy importance $\tilde{RI}_j$ of each design of conservation can be calculated applying equation (3).

$$\tilde{RI}_j = \sum_{i=1}^{n} \tilde{W}_i^* \otimes \tilde{R}_{ij}, \ j = 1, 2, ..., m \qquad (3)$$

Where $\tilde{W}_i^*$ is the fuzzy weighted importance of *i*th dimension of ecotourism, while $\tilde{R}_{ij}$ is the fuzzy number expressing the impact of the *j*th *CD* versus the *i*th *EA*.

In a similar manner, the position $\tilde{T}_{kj}$, $j, k = 1, 2,..., m$, $k \neq j$, in the correlations matrix expresses the fuzzy correlation between the *k*th and *j*th "hows". In order to quantitatively ponder the fuzzy correlation between "hows", we adopt the $\tilde{T}_{kj}$ which can be interpreted as the incremented changes of the degree of attainment of the *j*th "how" when the attainment of the *k*th one is unitary increased. Thus, the fuzzy weighted importance $\tilde{RI}_j^*$ can be computed as equation (4).

$$\widetilde{RI}_j^* = \widetilde{RI}_j \oplus \sum_{k=j} \widetilde{T}_{kj} \otimes \widetilde{RI}_k, \quad j = 1, 2, ..., m \tag{4}$$

2.3. Fuzzy Cost and Marginal Benefit of the Design of Conservation

In order to complete the assessment of the design of conservation, their cost of implementation should be considered. In this situation, fuzzy logic becomes a fundamental tool in dealing with ill-defined issues such as the evaluation of costs. While the national park authority find objective difficulties in quantitatively assessing the costs of implementation of the design of conservation, planners or project managers of the national park can more easily give a judgement on a linguistic scale, ranging for instance from *Very High* to *Very Low*. That is why, in the lower part of the HoQ, a fuzzy parameter $\widetilde{C}_j$ has been added to consider the cost of implementing the *j*th the design of conservation.

The fuzzy marginal benefit $\widetilde{U}_j$ of the design of conservation can be calculated through the ratio between benefits and costs, as expressed by the equation (5).

$$\widetilde{U}_j = \widetilde{RI}_j^* \otimes \frac{1}{\widetilde{C}_j}, \quad j = 1, 2, ..., m. \tag{5}$$

Since both $\widetilde{RI}_j^*$ and $\widetilde{C}_j$ parameters are fuzzy numbers, equation (5) describes an operation between fuzzy numbers and the resulting $\widetilde{U}_j$ is thus a fuzzy number. In order to understand local residents' degree of satisfaction and make *CDs* comparable results, normalizing and retranslating the fuzzy numbers into linguistic terms should be described.

Suppose that the fuzzy number, $\widetilde{B} = (b_1, b_2, b_3)$, can be normalized by diveded its upper bounds, b_3, which means the fuzzy most marginal benefit, and furthermore, transform the normalized fuzzy number into the crisp number by equation (6) (Yager, 1981).

$$x = \frac{b_1 + 2b_2 + b_3}{4} \tag{6}$$

The linguistic term, *A*, can be represented as the fuzzy number $\widetilde{A} = (a_1, a_2, a_3)$, where $a_1 \leqq a_2 \leqq a_3$. The fuzzy number $\widetilde{B}$ is "*approximately the linguistic term A*", has the numbership function :

$$u_{\widetilde{A}}(x) = \begin{cases} 0 & , \quad x < a_1 \text{ or } x > a_3 \\ (x - a_1)/(a_2 - a_1) & , \quad a_1 \leq x \leq a_2 \\ (a_3 - x)/(a_3 - a_2) & , \quad a_2 < x \leq a_3 \end{cases}$$

Where x is the crisp value transformed by equation (6). Thus, $u_{\widetilde{A}}(x)$ represents the possibility that the fuzzy number $\widetilde{B}$ is "approximately the linguistic term A".

Suppose that the fuzzy set, $\widetilde{A} = \{\sum_{i=1}^{n} u_{\widetilde{A}_i}(x) / A_i\}$, represents the possibility that the fuzzy number, $\widetilde{B}$, is "approximately the linguistic terms $A_1, A_2, ..., A_n$". The triangular fuzzy number $\widetilde{B}$ can be translated into the linguistic terms A_k, where $1 \leq k \leq n$, as equation (7):

$$\frac{u_{\widetilde{A}_k}(x)}{A_k} = \max\{\sum_{i=1}^{n} \frac{u_{\widetilde{A}_i}(x)}{A_i}\} \qquad (7)$$

3. Empirical Study: Kinmen National Park

In this paper, the methodology developed is applied to a ecotourism development, which refers to a major environmental operating in the design of conservation of Kinmen national park. The main objective of the application is twofold. On the one hand, it is aimed at assessing its robustness and consistency, where robustness and consistency are respectively understood to be related to the applicability of the methodology and to the reliability of the result obtained. On the other hand, the application strives to consider practical implications in managing guidelines and principles of ecotourism development a FQFD approach.

3.1. Research Setting in Kinmen

Kinmen national park, the sixth national park in Taiwan, is involved in ecological conservation and protection of the natural landscape, in addition to historically and culturally important sites and battlefield monuments. Kinmen is located off the southeastern coast of Fujian Province in Xiamen Bay at the outlet to the Jiulong River. This area includes Kinmen Island, Liehyu (also known as Little Kinmen), DaDan, and ErDan, of a total of twelve islands and islets. These islands and islets cover an area of 150 square kilometers. To the west, at a distance of about 10 kilometers, is the Xiamen outport. To the east, at a distance of 277 kilometers, is Taiwan (Fig. 2).

Kinmen national park is home to many endangered species, such as Eurasian otter, magpie robin, blue-tailed bee eater, lesser pied kingfisher, black-collared starling, black-winged hawk, Dicrurus hottentottus, Litsea glutinosa, Pyrus betulifolia, Abelia chowii Hoo and the evening primrose species Oenothera drummondii, etc (Forestry Bureau, 2007). From survey data and a review of the literature, Kinmen National Park is home to at least 8 species of mammals, 283 species of birds, 13 species of reptiles, 5 species of amphibians, 45 species of butterflies, 32 species of mollusks and 6 species of crustaceans (Kinmen County Government, 2007).

Kinmen is not only rich in its biodiversity but also in the cultural heritage of recording the footsteps and the endeavors of Chinese ancestors over the past hundreds of years. Most of the early inhabitants came from the Zhangzhou and Quanzhou areas of Fujian Province. The architectural style of the dwellings and local customs follow in the old traditions. Forty years of military control have slowed the pace of Kinmen's urbanization, enabling its historical heritage to be preserved. Thus, it can be said that Kinmen National Park possesses a rich culture and history, expressed in historical sites and traditional architecture. Inside the park are 11 registered historical monuments. Traditional villages and architecture are the richest cultural assets of Kinmen National Park. The seven representative traditional settlements of Oucuo, Jhushan, Shueitou, Cyonglin, Shanhou, Nanshan and Beishan have mostly retained their southern Fujian architecture in the Zhangzhou and Quanzhou styles.

KinSha and ShanHo, the villages under study, are two primary communities at the periphery of the Kinmen national park. KinSha is suited next to the check point of the Kinsha Reservoir and ShanHo little further away (see Fig. 2). Zhangzhou and Quanzhou people make up 75% of the resident population in KinSha and 80% in ShanHo. Since the 1980s, tourism has become one of the main economic and social activities in KinSha and ShanHo. Out-migration combined with aging of the resident population, has compelled the local government to search for better livelihood opportunities. Hence, a tourism development project in and around the Yangshan Bay, which is located within the reserve near the Kinsha Reservoir, was proposed to help augment the local economy.

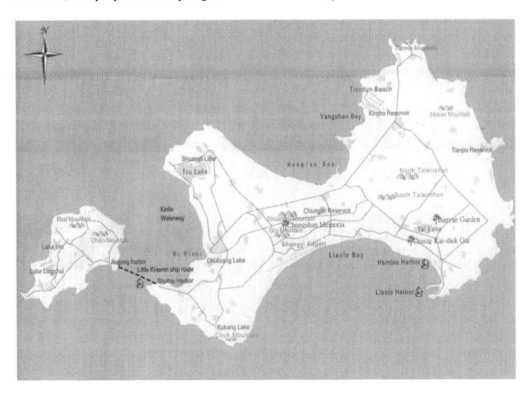

Figure 2. Kinmen national park showing the natural ecology.

Yangshan Bay has been a tourist attraction since 1980s. After the establishment of the reserve, tourist activities were declared illegal but have not been completely stopped. The unique and pristine environment of the area as well as the mythical meanings attached to it as a sacred place for the Zhangzhou and Quanzhou people have invited illegal visits to continue, especially between November and March when the water level of Tiandun Beach remains low. Littering and other ecological impacts caused by tourist activities became a concern to the local communities, especially to tribal group of Zhangzhou and Quanzhou now residing in KinSha, whose traditional territory once included the area where the reserve is located. The proposed ecotourism project was conceived not only as an economic catalyst but also as a way to strengthen conservation by involving and empowering local people in then management of the Kinmen national park.

3.2. Procedure

Fieldwork for this study was conducted in the year 2007. According to the Kinmen County Government census data of 2007, KinSha consisted of 1163 residents and 286 households, and ShanHo 589 residents and 143 households. Most of the local residents between the ages of 20 and 55 were either studying or working outside the villages and returned only occasionally. Therefore, a quota sampling of local residents was implemented to equally represent different age and gender groups from both communities. The sample size for each community was proportional to its population; therefore, the number of respondents in KinSha was two times that of ShanHo.

The snowball method was applied to interview local residents, and interviews were guided by a survey questionnaire. In total, 75 face-to-face interviews were completed, 49 in KinSha and 26 in ShanHo. To overcome the language barrier, a local resident volunteered to help in interviews with some respondents, mostly over 50 yeas old, who could not read Chinese characters or did not speak fluent Mandarin, the official language of Taiwan, since they grew up speaking their native language. Besides, in order to evaluate various designs of conservation of Kinmen national park and the relationship between dimensions of ecotourism and guidelines of ecotourism design/management requirements, eight officials of the national park come from Kinmen County Government were visited to request for filling the questionnaire.

Three measurement scales, included in the questionnaire, were developed to explore local residents' attitudes and local officials' intentions toward ecotourism development. Firstly, the information of "WHAT" in QFD comes from local residents; thus, those who offer "WHAT" information should realize residents' needs on ecotourism development. After the local people reveal their needs for the ecotourism development, the planners or project officials of the national park should develop a set of "HOW"s to capture the local members' needs in measurable and operable technical terms. Finally, this is an important work in HoQ/QFD which is performed carefully and collectively by researchers. The relationship between the design of conservation (HOW) and the requirements of ecotourism development (WHAT) is usually determined by analyzing to what extent the HOW could technically relate to and influence the WHAT. The data were analyzed by fuzzy linguistic scale responses to there categories to facilitate data interpretation. The fuzzy linguistic attitudes about ecotourism development ranked from very unfavorable to very favorable, and the intentions about supportive behaviors of ecotourism reflected the likelihood of engaging distinguished from highly unlikely to highly likely.

4. Results and Discussion

4.1. Identifying the Dimensions of Ecotourism Development and the Designs of Conservation

This study adopts four general dimensions identified from selected literature of ecotourism guidelines and principles to encompass the areas of socially appropriate tourism (Cooke, 1982), environmentally sustainable tourism (Wight, 1994), ecotourism (Honey, 1999; Wallace, 1996), and community based ecotourism (Sproule & Suhandi, 1998). The three dimensions identified include conservation of natural resources, preservation of cultural tradition, sustainable community development.

When applying the proposed HoQ to the ecotourism development, appropriate "whats" have to be identified. The main ecotourism development attributes "whats" to be considered in the Kinmen national park application have emerged from a preliminary survey phase, which has been performed through direct interviews carried out by academicians with 75 local residnets visted in KinSha and ShanHo. The relevant ecotourism development attributes "whats" are shown in Table 1, together with a brief description.

Table 1. List of viable indicators for the dimensions of ecotourism development

Dimensions of Ecotourism	Ecotourism Development Attributes	Description
Dimension 1: conservation of natural resources	Capacity	Specifying carry in capacity for tourist activities in Kinmen national park reserve.
	Reserve Regulations	Relaxing the reserve regulations to facilitate tourism development
	Natural Heritage	Learning about the natural heritage of the area.
Dimension 2: preservation of cultural tradition	Traditional Ceremonies	Preserving the spirit and content of the traditional ceremonies from any change induced by tourism development.
	Original Economic	Replacing the original economic activities by tourism.
Dimension 3: sustainable community development	Crowds	Crowds of tourist in the community.
	Investment	Maximizing non-local tourism investment.
	Tourism Impacts	Preventing negative tourism impacts.

The second part of this study is to establish deliver process of the design of conservation by the depth-interview with eight officials of the national park come from Kinmen County Government to collect relational variables. The key question to ask in this step is "how" the national park authority delivers the design of conservation. A list of possible "hows" when local members' needs related to ecotourism design/management requirements have to be improved is shown in Table 2. Another, the eight officials of the national park added four new conservation designs "hows" were identified based both on literature analysis and the ecotourism development characteristics, whose peculiarities have emerged from round-table discussions. Results are shown in Table 2 with a brief description for each original and new added design of conservation.

Table 2. The design of conservation considered in the national park

Conservation delivers		Conservation designs	Description
Conservation of natural resources	Original	Regulation	Encouraging the management authority to have no regulation on tourist number in Kinmen national park reserve.
		Illegal Activities	Assisting reserve managers to prevent illegal activities.
	New	Relaxation of Regulations	Encoruaging the relaxation of the reserve regulations for tourism development.
Preservation of cultural tradition	Original	Traditional Events	Suggesting the local government to reschedule the traditional events to attract more tourists.
		Economic Activities	Encouraging the local government to replace the original economic activities by tourism.
	New	Cultural Heritage	Learning about the cultural heritage of the area.
Sustainable community development	Original	Environmental Education	Providing environmental education for tourists.
		Crowds of Tourists	Welcoming crowds of tourists to the community regardless how many of them.
		Tourist Destination	Suggesting the local government to develop popular tourist destination.
	New	Non-local Tourism Investment	Encouraging the local government to maximize non-local tourism investment
		Negative Tourism Impacts	Involving in the prevention of negative tourism impacts

The third part of application focused on the assessment of viable conservation designs "hows", their mutual correlations, as well as of the relationships judgments between ecotourism attributes and the design of conservation. The planners or project officials of the national park adopt a linguistic approach; therefore, a first instructive phase was required to introduce the conservation design members to fuzzy set concept. In a similar manner, appropriate linguistic scales were set up for the evaluation of relative and weighted importance of *SAs*, the relative and weighted importance of *CDs*, together with values in the relationships and correlations matrixes.

4.2. Relative Importance Ratings in the Ecotourism Requirements

4.2.1. Fuzzy Linguistic Spreads of Dimensions of Ecotourism Development

During the survey phase, the 75 local residents have also been asked about the favor of ecotourism attributes, in order to determine the relative favor of ecotourism attributes, as well as to assess the distance between the ecotourism delivered for each attribute and the importance that is perceived as surperior. The 75 local residents have been asked to rank the favor attitude of each ecotourism attribute on a 9-point linguistic rating scale, ranging from *VL (Very Low)* to *VH (Very High)*. The fuzzy scale is in Fig. 3.

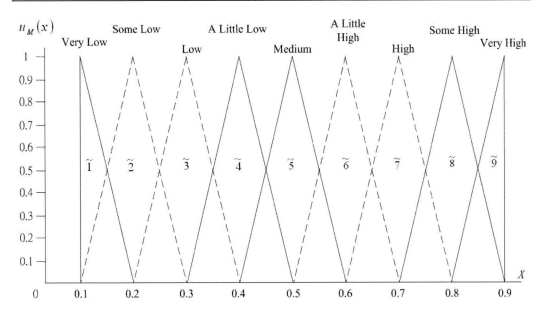

Figure 3. The local residents' linguistic importance terms.

$\widetilde{w}_{i,x}$ is the fuzzy triangular number which is adopted to translate the linguistic favor judgement given to the ith EA by the xth local resident, $\widetilde{w}_{i,x}$ fuzzy numbers have been pooled to determine an aggregate value to be used in the HoQ, which is the relative favor attitude $\widetilde{W}_i$ previously defined. To this extent, the relative favor attitude $\widetilde{W}_i$ of ecotourism attribute ith can be computed as a favor attitude average of $\widetilde{w}_{i,x}$. The favor attitude average takes into account the issue that not all local residents are equal. In the specific case, the following equation is applied.

$$\widetilde{W}_i = \sum_{x=1}^{75} I_x \otimes \widetilde{w}_{i,x}, \quad i = 1, 2, \ldots, n. \tag{8}$$

I_x is the favor attitude of xth local resident surveyed ($x = 1, 2, \ldots, 75$).

The importance of each local resident has been weighted through the percentage of staying Kinmen, as shown in Table 3. The resulting respondents of sample consisted of 43 men (57.3%) and 32 women (42.7%), where the average was 43.5 years old.

Table 3. Importance ranking of the local residents of Kinmen national park

Local resident	Staying Years	Importance	Importance judgment
L_1	79	0.095	Very High
L_2	77	0.085	Very High
L_3	76	0.080	Some High
⋮	⋮	⋮	⋮
L_{75}	26	0.015	Very Low

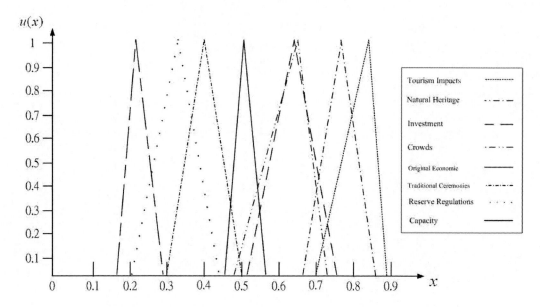

Figure 4. The spread of fuzzy number among the ecotourism attributes.

Based on values shown in Table 3, the importance of each local resident has been expressed with a fuzzy importance judgment using the same 9-point linguistic scale. The resulting fuzzy numbers have been used in the computation of $\widetilde{W}_i$, and results are shown in Table 4. The triangular have relatively heavy area in "Natural Heritage" and "Traditional Ceremonies" compared with the spread of fuzzy number among the ecotourism attributes in Figure 4. This shows local residents have more recognition discrepancy in these two ecotourism attributes. Besides, the triangular areas in ecotourism attribute of "Capacity" and "Original Economic" are relatively small which means local residents with recognition relatively unanimous and more consistently. Furthermore, the triangular area is more similar in "Investment" and "Crowds" with overlap the area quite largely. This shows that the preference difference is not large between the two ecotourism attributes. By all accounts, the local residents consider Tourism Impacts, Natural Heritage, Investment and Crowds the important ecotourism attributes.

4.2.2. Fuzzy Weighted Importance by Local Residents' Cognition Gaps

Once $\widetilde{W}_i$ were calculated, the weighted preference $\widetilde{W}_i^*[n \times 1]$ of *EAs* was computed in according with equation (2). As regards to the crisp distance d_i between the ecotourism attributes' performance and the one that is perceived by local residents as superior, the parameter has been computed as the average of crisp distances $d_{i,x}$, the generic xth local redident perceives against ith ecotourism attribute, as shown in equation (9):

$$d_i = \frac{\sum_{x=1}^{75} d_{i,x}}{75} , \ i = 1, 2, \ldots, n. \tag{9}$$

Table 4. Fuzzy importance $\widetilde{W}_{i,x}$ assigned to ecotourism attributes by local residents and the relative importance of ecotourism attributes $\widetilde{W}_i$

Ecotourism attributes		Capacity	Reserve Regulations	Natural Heritage	Traditional Ceremonies	Original Economic	Crowds	Investment	Tourism Impacts
Importance judgment	L_1	L	L	SH	L	M	AH	H	VH
	L_2	SL	SL	VH	M	AH	H	SH	VH
	L_3	AL	M	AH	AL	AL	H	SH	SH
	...	...	...	...	...	...	...	...	...
	L_{75}	VL	L	AH	M	M	VH	VH	H
Relative Importance $\widetilde{w}_{i,x}$	L_1	(0.2,0.3,0.4)	(0.2,0.3,0.4)	(0.7,0.8,0.9)	(0.2,0.3,0.4)	(0.4,0.5,0.6)	(0.5,0.6,0.7)	(0.6,0.7,0.8)	(0.8,0.9,0.9)
	L_2	(0.1,0.2,0.3)	(0.1,0.2,0.3)	(0.8,0.9,0.9)	(0.4,0.5,0.6)	(0.5,0.6,0.7)	(0.6,0.7,0.8)	(0.7,0.8,0.9)	(0.8,0.9,0.9)
	L_3	(0.3,0.4,0.5)	(0.4,0.5,0.6)	(0.5,0.6,0.7)	(0.3,0.4,0.5)	(0.3,0.4,0.5)	(0.6,0.7,0.8)	(0.7,0.8,0.9)	(0.7,0.8,0.9)
	...	...	...	...	...	...	...	...	...
	L_{75}	(0.1,0.1,0.2)	(0.2,0.3,0.4)	(0.5,0.6,0.7)	(0.4,0.5,0.6)	(0.4,0.5,0.6)	(0.8,0.9,0.9)	(0.8,0.9,0.9)	(0.6,0.7,0.8)
Relative Importance of Ecotourism Attributes $\widetilde{W}_i$		(0.19, 0.21, 0.28)	(0.22, 0.31, 0.43)	(0.67, 0.79, 0.86)	(0.30, 0.40, 0.50)	(0.40, 0.50, 0.55)	(0.48, 0.65, 0.73)	(0.52, 0.61, 0.75)	(0.70, 0.85, 0.88)

Table 5. Distance d_i from the optimum performance and weighted importance $\widetilde{W}_i^*$ of each ecotourism attribute

	Performance judgments					Optimum performance					Distance d_{ix}					Distance d_i	Relative importance	Weighted importance
	L_1	L_2	L_3	...	L_{75}	L_1	L_2	L_3	...	L_{75}	L_1	L_2	L_3	...	L_{75}			
Capacity	H	SH	AH	...	H	H	VH	SH	...	H	0	0.1	0.15	...	0	0.072	(0.19, 0.21, 0.28)	(0.013, 0.015, 0.020)
Reserve Regulations	M	AH	H	...	AH	M	H	H	...	VH	0.1	0.1	0	...	0.2	0.093	(0.22, 0.31, 0.43)	(0.020, 0.028, 0.040)
Natural Heritage	AL	M	M	...	L	H	SH	H	...	SH	0.2	0.15	0.15	...	0.3	0.371	(0.67, 0.79, 0.86)	(0.248, 0.293, 0.319)
Traditional Ceremonies	M	AH	AL	...	H	AH	AH	H	...	SH	0.2	0	0.2	...	0.1	0.284	(0.30, 0.40, 0.50)	(0.085, 0.113, 0.142)
Original Economic	H	AH	SH	...	AH	H	H	SH	...	AH	0	0.1	0	...	0	0.049	(0.40, 0.50, 0.55)	(0.019, 0.024, 0.027)
Crowds	AH	AH	M	...	AH	AH	AH	AH	...	VH	0	0.1	0.1	...	0	0.164	(0.48, 0.65, 0.73)	(0.078, 0.11, 0.12)
Investment	AH	H	M	...	H	AH	H	H	...	VH	0	0	0.15	...	0	0.525	(0.52, 0.61, 0.75)	(0.273, 0.32, 0.394)
Tourism Impacts	AL	SL	M	...	L	H	H	SH	...	SH	0.2	0.3	0.25	...	0.35	0.358	(0.70, 0.85, 0.88)	(0.251, 0.304, 0.315)

Parameters $d_{i,x}$ have been obtained based on the survey results and by applying equation (1). To this extent, a section of the survey was dedicated to importance judgments about the ecotourism attribute delivered by the national park authority to its local members. The local residents were asked to judge the ecotourism development level they were receiving for each ecotourism attribute, using the linguistic scale shown in Fig. 3. Moreover, for each *EA*, the local residents had to indicate the judgment which best matched their perception of a superior ecotourism attribute. $d_{i,x}$ parameters, d_i values, and the corresponding weighted preference $\widetilde{W}_i^*$ are shown in Table 5. From outcomes analysis, it emerges that local residents perceive a significant difference between the ecotourism attribute performance and optimum one in terms of "Traditional Ceremonies", "Natural Heritage", "Investment" and "Tourism Impacts". As can be seen comparing Table 4 and Table 5, "Tourism Impacts", "Investment" and "Natural Heritage" are considered the most three important ecotourism attributes from local residents' point of view, but since the performance delivered is far from meeting local people requirements, they should be considered the key ecotourism attributes to tune.

4.3. Fuzzy Relationship Matrix Assessment

4.3.1. Fuzzy Comparisons Matrices between Ecotourism Attributes and Conservation Designs

This step in the construction of the HoQ was the assessment of the fuzzy relationship matrix $\widetilde{R}_{i,j}[n \times m]$. To this extent, the design of conservation *CDs* for local residents satisfaction have been listed in columns, while ecotourism attributes *EAs* have been crossed over in rows. The degree of relationship (weak, medium, strong) between *CDs* and *EAs* has been expressed by eight project officials of the national park using linguistics cognitions, which one usually adopted in crisp QFD approaches. Since fuzzy logic is exploited to well cope with the ill-defined nature of linguistics judgments, graphics symbols and linguistics should have been translated into many fuzzy triangular numbers instead of crisp ones. Table 6 shows the correspondence between symbols and fuzzy numbers.

Table 6. Degree of relationship, graphic symbols and corresponding fuzzy numbers

Degree of relationship	Graphic symbol	Fuzzy number
Strong	●	(0.7,0.8,0.9)
Medium	○	(0.4,0.5,0.6)
Weak	▲	(0.1,0.2,0.3)

The resulting relationship matrix between ecotourism attributes and the design of conservation from eight project officials' linguistic cognitions is shown in the centre of Table 7.

Table 7. The relationship between ecotourism attributes and conservation design in house of quality

		Conservation designs										
		Regulation	Illegal Activities	Relaxation of Regulations	Traditional Events	Economic Activities	Cultural Heritage	Environmental Education	Crowds of Tourists	Tourist Destination	Non-local Tourism Investment	Negative Tourism Impacts
Ecotourism Attributes	Capacity	Medium		Strong		Strong	Medium		Strong			Weak
	Reserve Regulations	Medium	Weak	Strong		Medium	Medium		Strong			Weak
	Natural Heritage	Weak	Medium	Strong	Weak		Weak	Weak	Medium			
	Traditional Ceremonies	Strong	Strong	Medium		Strong	Medium		Medium			
	Original Economic	Strong	Medium	Medium			Weak		Weak			
	Crowds				Strong			Medium		Strong	Strong	Weak
	Investment				Strong			Strong		Strong	Medium	Strong
	Tourism Impacts	Weak		Medium	Weak			Strong		Medium	Medium	Strong

4.3.2. Fuzzy Comparisons Matrices of Interdependencies with the Designs of Conservation

The roof of correlations was built up in a similar manner. Traditional QFD linguistics have been used to express the correlations between conservation designs (strong negative, negative, positive, strong positive) and symbols have been translated into fuzzy numbers, as shown in Table 8.

Table 8. Degree of correlation, graphic symbols and corresponding fuzzy numbers

Degree of correlation	Graphic symbol	Fuzzy number
Strong positive	●	(0.7,0.8,0.9)
Positive	○	(0.5,0.6,0.7)
Negative	□	(0.3,0.4,0.5)
Strong negative	■	(0.1,0.2,0.3)

Once the relationship matrix and the roof of correlations were compiled, the fuzzy relative importance $\widetilde{RI}_j$ and the fuzzy weighted importance $\widetilde{RI}_j^*$ of each conservation design were computed in accordance with equation (3) and (4) respectively. Results are shown in Fig. 5.

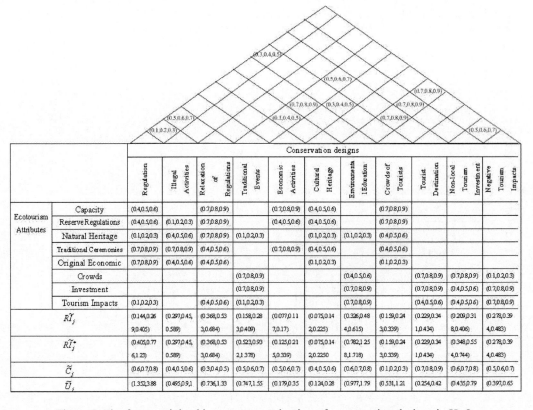

Figure 5. The fuzzy weighted importance evaluation of conservation designs in HoQ.

4.4. Final Importance Ratings of the Design of Conservation

Then, the fuzzy cost $\tilde{C}_j$ for the implementation of each conservation design was determined to evaluate the fuzzy marginal benefit $\tilde{U}_j$. To this extent, the eight project officials were asked to express a linguistic judgment about the investment required for each conservation design, by using the 9 value fuzzy scale previously shown in Fig .3. Results are shown in Fig. 5. It should be remarked that fuzzy logic was found to be a very flexible tool to handle such a vague, imprecise and ill-defined issue as costs estimation for conservation designs.

From Fig. 4, the "Tourist Destination" emerged as the conservation action with the highest implementation priority, since, despite the very high cost for implementation, it makes it possible to improve the important conservation attributes, such as "Crowds" and "Investment". In addition, the "Tourist Destination" has positive interdependence against other conservation designs. In particular, a strong positive relationship can be found between the "Environmental Education" implementation and there is a positive relationship between the "Traditional Events".

4.5. The Change of Linguisctic Satisfaction in New Conservation Designs

The fuzzy resulting benefits $\tilde{U}_j$ have been computed according to equation (5). Finally, fuzzy $\tilde{U}_j$ parameters were normalized by dividing its upper bounds of fuzzy marginal benefits and defuzzified applying equation (6). Crisp U_j obtained can be regarded as synthesis parameters, expressing the overall local residents' satisfaction index of implementing the *j*th conservation design. Thus, based on equation (7), the crisp value was translated into the linguistic term as shown in Fig. 6. For example, the satisfaction index of "Negative Tourism Impacts" (0.692) that is approximately two linguistic variables, "satisfied" and "a little satisfied", is computed by membership function as:

For "satisfied", $u_{satisfied}(x_{price}) = 0.089$

For "a little satisfied", $u_{a\ little\ satisfied}(x_{price}) = 0.93$

Also by mapping the crisp value of "Negative Tourism Impacts" to membership function, the local resident satisfaction index is approximately the other linguistic terms (Fig. 6) is zero. Finally, based on equation (7), the satisfaction index of "Negative Tourism Impacts" is assigned the linguistic term "a little satisfied".

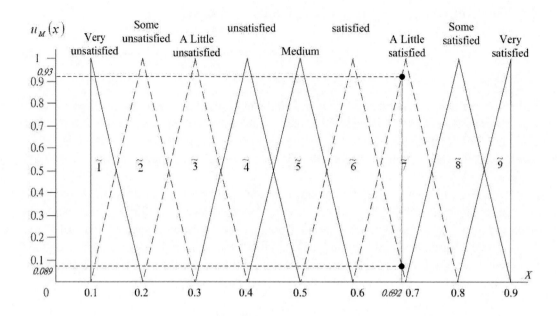

Figure 6. The local residents' linguistic satisfaction terms.

The linguistic satisfaction index of original and new added conservation design alternatives of the ecotourism development can be shown in Fig. 7. As a result, new added conservation designs are quantitatively evaluated by the fuzzy quality function deployment management. The local resident satisfaction index of the new conservation design is higher than the local resident satisfaction index of the existing conservation designs of Kinmen national park in Fig. 7. For example in the "Conservation of natural resources", the original conservation designs include Regulation and Illegal Activities, as to local residents, the linguistic spreads of satisfaction to conservation designs are "some unsatisfied" and "unsatisfied" respectively. The local residents have the extent difference for "Regulation" from "some unsatisfied" to "some satisfied" after adding the new conservation designs, "Relaxation of Regulations". Besides, the "Illegal Activities" is promoted from "unsatisfied" to "a little satisfied" and "Relaxation of Regulations" presents the state of "satisfied", too. Therefore, in the items of original conservation designs, the whole satisfaction preference of local residents had a partiality for changed from "unsatisfied" into "some satisfied" when the new conservation designs added, namely local resident satisfaction index.

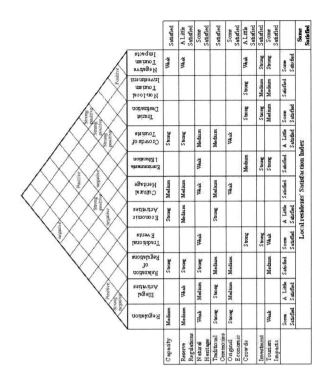

Figure 7. FQFD for evaluating conservation design alternatives.

5. Conclusion

This study has addressed the applicability of FQFD in the dimensions of ecotourism and guidelines of ecotourism design/management requirements. The proposed methodology developed could be rightly considered as a useful tool for selecting the efficient and effective ecotourism development to reach local people satisfaction. In particular, the methodology allows the identification of dimensions of ecotourism attributes that are perceived to affect conservation designs performances from the local residents' point of view, enabling the assessment of possible gaps between local residents and the management authority of Kinmen national park with the perception of ecotourism development. As a matter of fact, this is why the perception of the authority should not be considered the starting point in developing ecotourism strategies, while direct interviews with local residents are required. In our approach, such an issue is addressed through the computation of the distance between ecotourism development performance in terms of ecotourism attributes and that which is perceived by local residents as superior.

Basing on the importance of local residents, the weighted importance of ecotourism attributes allows the national park authority to identify the key factors of intervention in order to improve the perceived ecotourism development. For example, "Tourism Impacts" emerges as the most important ecotourism attribute from local residents' point of view, while, based on the distance judgments the local residents give on the performance and optimum, then, "Investment" and "Natural Heritage" should be considered the key ecotourism attributes to tune.

In order to assess viable conservation designs, in the approach proposed we have introduced a utility factor, which considers the costs of implementation for each "how". The utility factor can be directly adopted as a synthesis parameter to select the suitable conservation action to implement. In addition, the FQFD approach proposed has made it possible to appraise the beneficial impact of strategic leverages over dimensions of ecotourism attributes in the ecotourism development as well as the positive correlations with other conservation designs.

Since personal judgments are required when building the conservation designs of Kinmen national park in HoQ, fuzzy logic has been adopted as a useful tool. Rationships and correlations have been appropriately translated into triangular fuzzy numbers through local residents and project officials weight fuzzy logic linguistic judgments. Moreover, fuzzy logic has allowed coping well with uncertainties and incomplete understanding of the relationships between "hows" and between "hows" and "whats". In addition, fuzzy logic becomes fundamental to dealing with several parameters that seem difficult to express in a quantitative measure. For example, detailed information about costs of implementation for conservation designs are usually not available, while linguistic judgments on costs can be easily obtained.

Finally, it can utilize the concept of fuzzy linguistic inference to compare the original and new added conservation designs that the management authority offered in the transition situation of local residents' cognitive satisfaction. Through measuring local residents' perceptions of ecotourism development in the authority of national park allows managers to better tailor marketing efforts to ensure that local residents' needs are met. As a result, the management authority of Kinmen national park can recognize, prioritize and improve areas of ecotourism development weakness and allocate important resources to the most effective conservation areas. Thus, the results from this research may have some significant implications for managers of ecotourism.

References

[1] Ajzen, I., & Fishbein, M. (1980). *Understanding attitudes and predicting social behavior*. Englewood Cliffs, NJ: Prentice-Hall.

[2] Allen, L. R., Long, P. T., Perdue, R. R., & Kieselbach, S. (1988). The impact of tourism development on citizen's perceptions of community life. *Journal of Travel Research*, **27**(1), 16-21.

[3] Alpert, P. (1996). Integrated conservation and development projects. *Bioscience*, **46**, 845-855.

[4] Bookbinder, M. P., Dinerstein, E., Rijal, A., Cauley, H., & Rajouria, A. (1998). Ecotourism's support of biodiversity conservation. *Conservation Biology*, **12**(6), 1399-1404.

[5] Brandon, K. (2001). Moving beyond Integrated Conservation and Development Projects (ICDPs) to achieve biodiversity conservation. In D. R. Lee, & C. B. Barrett (Eds.), *Tradeoffs or synergies? Agricultural intensification, economic development and the environment* (pp. 417–432). New York, NY: CABI Publishing.

[6] Brandon, K., & Margoluis, R. (1996). *Structuring ecotourism success: Framework for analysis*. In E. Malek-Zadeh (Ed.), The ecotourism equation: Measuring the impacts (pp. 28-38). New Haven, CT: Yale University School of Forestry and Environmental Studies.

[7] Chapman, D. (2003). Management of national parks in developing countries—A proposal for an international park service. *Ecological Economics*, **46**, 1-7.

[8] Chi, C. C., & Wang, J. J. (1996). *An analysis of the conflicts between the indigenous peoples and national parks in Taiwan*. Retrieved 9 October, 2005 from the World Wide Web (10/08/05): http://wildmic.npust.edu.tw/sasala/

[9] Chien, C. J., & Tsai, H. H. (2000). Using fuzzy numbers to evaluate perceived service quality. *Fuzzy Sets and Systems*, **116**, pp. 289-300.

[10] Cooke, K. (1982). Guidelines for socially appropriate tourism development in British Columbia. *Journal of Travel Research*, **21**(1), 22-28.

[11] Eagly, A. H., & Chaiken, D. (1993). *The psychology of attitude*. New York, NY: Harcourt Brace Jovanovich, Inc.

[12] Fishbein, M., & Manfredo, M. J. (1992). A theory of behavior change. In M. J. Manfredo (Ed.), *Influencing human behavior: Theory and applications in recreation, tourism, and natural resource management* (pp. 29-50). Champaign, IL: Sagamore.

[13] Forestry Bureau (2007). *Nature Reserve*. Retrieved 2 May, 2007, from the World Wide Web: http://163.29.26.177/conservation-2-2-6all.html.

[14] Getz, G. (1994). Residents' attitudes toward tourism: A longitudinal study in Spey Valley, Scotland. *Tourism Management*, **15**(4), 247-258.

[15] Gibson, C. C., & Marks, S. A. (1995). Transforming rural hunters into conservationists: An assessment of community-based wildlife management programs in Africa. *World Development*, **23**, 941-957.

[16] Hackel, J. D. (1999). Community conservation and the future of Africa's wildlife. *Conservation Biology*, **13**(4), 726-734.

[17] Heinen, J. T. (1996). Human behavior, incentives and protected area management. *Conservation Biology*, **10**, 681-684.

[18] Honey, M. (1999). *Ecotourism and sustainable development: Who owns paradise?* Washington, DC: Island Press.

[19] Hsu, T. H., & Lin, L. Z. (2006). Using Fuzzy Set Theoretic Techniques to Analyze Travel Risk: An Empirical Study. *Tourism Management*, **27**, 968-981.

[20] Hsu, T. H., & Lin, L. Z. (2006). QFD with Fuzzy and Entropy Weight for Evaluating Retail Customer Values. *Total Quality Management*, **17** (7), 935-958.

[21] Liu, J., & Var, T. (1986). Resident attitudes toward tourism impacts in Hawaii. *Annals of Tourism Research*, **13**, 193-214.

[22] Long, P. T., Perdue, R. R., & Allen, L. (1990). Rural resident tourism perceptions and attitudes by community level of tourism. *Journal of Travel Research,* **28**(3), 3-9.

[23] Mason, P., & Cheyne, J. (2000). Residents' attitudes to proposed tourism development. *Annals of Tourism Research*, **27**(2), 391-411.

[24] Nepal, S. K. (2002). Involving indigenous peoples in protected area management: Comparative perspectives from Nepal, Thailand, and China. *Environmental Management*, **30**(6), 748-763.

[25] Nepal, S. K., & Weber, K. E. (1995). Managing resources and resolving conflicts—National parks and local people. *International Journal of Sustainable Development and World Ecology*, **2**, 11-25.

[26] Newmark, W. D., & Hough, J. L. (2000). Conserving wildlife in Africa: Integrated conservation development projects and beyond. *BioScience*, **50**(7), 585-592.

[27] Ryan, C. (2000). *Indigenous peoples and tourism*. In C. Ryan, & S. Page (Eds.), Tourism management: Towards the new millennium (pp. 421-430). Oxford: Pergamon.

[28] Smith, M., & Krannich, R. (1998). Tourism dependence and resident attitudes. *Annals of Tourism Research*, **25**, 783-802.

[29] Spiteri, A., & Nepal, S. K. (2006). Incentive-based approaches to conservation in developing countries. *Environmental Management*, **37**, 1-14.

[30] Sproule, K. W., & Suhandi, A. S. (1998). *Guidelines for community-based ecotourism programs —Lessons from Indonesia.* In K. Lindberg, M. Epler Wood, & D. Engeldrum D (Eds.), Ecotourism: A guide for planners and managers, Vol. 2 (pp. 215-235). North Bennington, VT: Ecotourism Society.

[31] Wallace, G. N. (1996). *Toward a principled evaluation of ecotourism venue.* In E. Malek-Zadeh (Ed.), The ecotourism equation: Measuring the impacts (pp. 119-140). New Haven, CT: Yale University School of Forestry and Environmental Studies.

[32] Walpole, M. J., & Goodwin, H. J. (2001). Local attitudes toward conservation and tourism around Komodo National Park, Indonesia. *Environmental Conservation*, **28**(2), 160-166.

[33] Wells, M. P., & Brandon, K. E. (1992). *People and parks: Linking protected area management with local communities.* Washington, DC: World Bank.

[34] Wells, M. P., & Brandon, K. E. (1993). The principles and practice of buffer zones and local participation in biodiversity conservation. *Ambio*, **22**(2-3), 157-162.

[35] Wight, P. A. (1994). *Environmental sustainable marketing of tourism.* In E. Cater, & G. Lowman (Eds.), Ecotourism: A sustainable option? (pp. 39-56). Brisbane: Wiley.

[36] Wunder, S. (2000). Ecotourism and economic incentives: An empirical approach. *Ecological Economics*, **32**(3), 465-479.

Chapter 5

PROTECTION OF NATURE – CHAMOIS IN THE SLOVAK NATIONAL PARKS

Štefančíková Astéria
Parasitological Institute Slovak Academy of Sciences,
Košice, Slovak Republic

Abstract

Slovak economy is influenced an expressive qualitative and quantitative changes nowadays. It used to be controlled by state exclusively, but at present is transformed into liberal market regime connected with the changes of the ownership and multiple problems to protection and exploitation of natural resources. A lot of plant and animal species disappeared, some became threatened as the result of expansive exploitation of natural resources. From total number of 548 free-living animals 153 are threatened. Tatras chamois (Rupicapra rupicapra tatrica Blahout, 1971) was included as a critically threatened species (IUCN Red List of Threatened Species) (Caprinae Specialist Group 2000). It is also classified as a threatened species by EMA (European Mammal Assessment) (Aulagnier et al. 2008). In the territory of Slovakia , the chamois is native only in High Tatras, West and Belianske Tatras. The fact, that the Tatras chamois are not only the glacial relict, but also the Tatras endemic subspecies (Blahout, 1971), considerably increases their cultural, historic and environmental importance. In order to prevent the chamois extinction, during 1969 – 1976, there was an attempt to create viable population of the Tatras chamois outside their original area. Therefore the Tatras chamois has been introduced to the central part of the Low Tatras National Park. The chamois of Alpine origin from Bohemia and Moravia has been introduced in the 60s into the national parks of Slovak Paradise and Veľká Fatra. The introduction of alpine chamois was hunting motivation of atractive species (Kratochvíl 1981). This paper analyses data concerning some aspects of the protection of nature in general, but predominantly concentrated on the chamois –their origin, ecological demands in the Slovak national parks, and also analyses the reasons of the depression the Tatras endemic subspecies in detail.

Introduction

Thanks to its geographical position in the middle of Europe and on the borderline of the Carpathians and the Pannonian basin, Slovakia is a home to an abundant diversity of flora and fauna. Biological diversity represents heterogeneity of all forms of life and affects ecosystem functions. In Slovakia, the first attempts to conserve wild nature and create conditions for the existence of protected areas supported by law date back to the period of feudal land ownership (the 13th -15th century) basically as protection of forests, free-living animals and medicinal springs. According to the Law of the Slovak National Council, Act. No 11/1948 Coll., the Tatras National Park was declared the first nature conservation area. Several years later according to the Law of the Slovak National Council, Act. No 1/1955 Coll., several protected areas of different category have gradually been declared, i.e. nine national parks, 14 monuments and other 899 protected, more or less isolated, areas. At present, nature conservation in Slovakia is regulated by the Law of the Slovak National Council Act. No. 287/1994 of the Coll. The new law has introduced the blanket concept of the nature conservation based on the territorial system of ecological balance, dividing the whole territoy into the five levels of protection and exploitation.

In the last decade, the Slovak economy has been undergoing major qualitative and quantitative changes. The economy, once controlled exclusively by the state, is heading in the direction of liberal market economy. This process is attended by the major changes in ownership and numerous problems regarding conservation and exploitation of natural resources. Many plant and animal species have dissapeared some became rare or threatened, as a result of over-exploitation of natural resources. Out of a total 548 free-living animal species, 153 are threatened.

Tatra mountain chamois (*Rupicapra rupicapra tatrica* Blahout, 1971) has been declared a critically threatened species by the IUCN Red List of Threatened Species (Caprinae Spedcialist Group 2000) and classified as a threatened species by the EMA (European Mammal Assessment) (Aulagnier et al. 2008). In the territory of Slovakia, the chamois is native only to the High Tatras, West and Belianske Tatras. The fact that the Tatra chamois are not only the glacial relict but also the Tatras endemic subspecies (Blahout, 1971) considerably increases their cultural, historic and environmental importance. In order to prevent the chamois extinction, during 1969 – 1976 there was an attempt to create viable population of the Tatra chamois outside their endemic area. The Tatra chamois has been introduced to the central part of the Low Tatras National Park. The chamois of an Alpine origin from Bohemia and Moravia has been succesfully introduced in the 1960s into the national parks of the Slovak Paradise and Veľká Fatra. The aim of such introduction was to provide a species attractive for hunting. As a result of tourism development and nutritional demands of the animals, the Alpine chamois population has expanded from its original areas of indroduction into the areas of the Tatra chamois potential occurrence. These facts have led to concerns about the possible crossbreeding that would threaten the original Tatra subspecies. The initial aim was to liquidate non-original chamois populations in Slovakia in order to maintain genetic identitiy and purity of the Tatra chamois in the central part of the Low Tatras National Park. Later, for the sake of maintaining biological diversity, a consensus has been reached to retain the Alpine chamois population and to conduct regulated shooting, in case of its observed migration near the biotopes of the Tatra chamois in the Low Tatras.

Characteristics of Biotopes

Tatra National Park (TANAP)

The Tatra National Park is the heritage of all mankind. In 1993 the Tatras mountain range, together with the Polish part of the National Park, was declared a natural UNESCO's biosphere reserve. The Tatra Mountains constitute the northernmost and the highest part of the 1200 km long Carpathian arch. Geologicaly young, they were folded during the late Tertiary period, i.e. 5-20 million years ago. As far as the distribution of chamois is concerned, the High Tatras and Belianske Tatras, differing in their geographical, orographical, geological, trophic and climatic conditions, are the most important sites. The geomorphological and orographical conditions are favourable with respect to chamois conservation. The habitats are situated at the subalpine up to alpine vegetation level with an altitude of 1749 – 2655 m above sea level (Fig. 1).

The Tatra Mountain range can be divided into two basic units - The West Tatras and The East Tatras. In the geological composition granites and crystalline schists prevail, especially in the orographic group of the High Tatras - in the East Tatras. In the Belianske Tatras -The East Tatras limestones and dolomites prevail. The current shape of the Tatra Mountains surface is a result of triple glaciation in the early Quaternary. Thus the characteristic relief features of the Tatra Mountains have originated, represented by peaks, glaciers, kettles (glacial cirques) and glacial troughs. The glacier recession was accompanied by the formation of moraine drifts and fluvioglacial cones at the Tatra Mountains base. The limestone parts of the Tatras are characterized by a frequent occurence of karst phenomena, such as karrens, abysses and limestone caves. Diversity of geological composition, surface relief, soil properties and climatic originality of the Tatras have given rise to fauna and flora of a special montane and even an Alpine character.

Above all in plant communities, there is a characteristic differentiation among five vegetation zones - submontane, montane, subalpine, alpine, nival. A similar diversity occurs in the animal kingdom. The Tatras has a continental climate with typical features of an Alpine climate, characterized by extreme temperature fluctuations.

Low Tatra National Park (NAPANT)

The Low Tatras National Park (NAPANT) is a National park in Central Slovakia, between the Váh River and the Hron River valleys. The park and its buffer zone cover the whole Low Tatras mountain range. The National Park covers an area of 728 km² and its buffer zone covers an area of 1,102 km², which makes it the largest national park in Slovakia. Several protected areas have been established within the area of the park or its buffer zone, covering 98.89 km²: 10 national nature reserves, 10 nature reserves, 5 national nature monuments, 6 nature monuments and one conservation site. The massif of the Low Tatras is the most important geographical and relatively high ecological unit of Slovakia. It differs from the High Tatras, because the activity of the glaciers was lower and typical summits (jokuls) occur less frequently than those in the High Tatras. Geological composition is a set of prevailing granites and crystalline schists, but limestones and dolomites are infrequent (Radúch and Karč 1981, Štefančíková, 1994, 2005). The habitats of chamois are situated at the subalpine to alpine vegetation level with an altitude of 1886 – 2003 m above sea level (Fig. 1). The climate is rather similar to that of the High Tatras (Houdek and Bohuš 1976).

Slovak Paradise National Park (NAPASR)

The Slovak Paradise is situated in the northwestern part of the Spiš Ore Mountains, which are a part of the Slovak Ore Mountains, belonging to the central zone of the Western Carpathians. Though not classified as high-mountains (400 – 1200 m above sea level), it is regarded as an area having an alpine character thanks to its well-known canyons with waterfalls and ravines. Chamois are distributed in the area south of Stratená up to the canyon of the Hornad River in the north and from Vernar up to Čingov in the west. It is a biotope with a mesozoic bedrock, consisting largely of Triassic limestones and dolomites. Chamois frequently occur in the localities that are situated at the altitude of 660 – 1057 m above sea level (Fig. l). The current population numbers app. 100 of chamois.

The climate is continental. While the northern part of the area is situated in the moderate temperature zone, the rest of the area lies in the cold zone and only tops and ridges have a warmer and drier climate determined by the rock composition of the mountains (Houdek and Bohuš 1976).

National Park Veľká Fatra (NAPAVF)

The National ParkVeľká Fatra and its protective zone comprise most of the Greater Fatra Range which belongs to the Outer Western Carpathians.The National Park was declared on April 1st. 2002 as an upgrade of the Protected Landscape Area of the same name established in 1972 in order to protect a mountain range with a high percentage of well-preserved Carpathian forests. National Park Veľká Fatra is also an important reservoir of freshwater thanks to high rain falls and low evaporation within the area. The core of the range is built of granite that reaches the surface only in certain places, more common are various slates creating gently modelled ridges and summits and limestone and dolomite rocks creating a rough and picturesque terrain. Many karst features, i.e. caves are also present there. Various rocks and therefore various soils, diverse type of terrain with gentle upland meadows and pastures, sharp cliffs and deep valleys provide for extremely rich flora and fauna. All species of large Central European carnivores are abundant there: brown bear, gray wolf and Eurasian lynx. The habitats of introduced alpine chamois are situated at the level with an altitude of 800 – 1350 m above sea level (Fig. 1). The current population numbers app. 53 chamois.

The climate is continental. Geographical position of Veľká Fatra mountains is the most important factor influencing the weather that reflects high climatic – temperature differences within the small territory (Houdek and Bohuš 1976). The climate is moderate cold and montane cold on the crests.

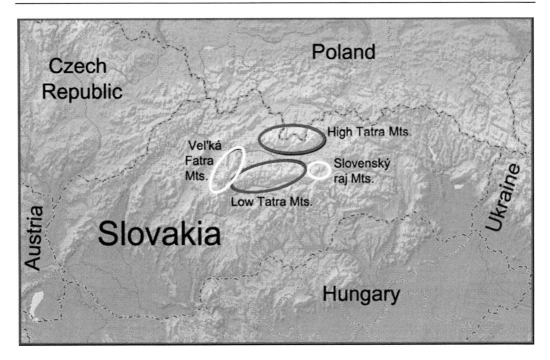

Figure 1. Mountain ranges of Slovakia inhabited by chamois (white – Alpine chamois, black – Tatra chamois (modified by Zemanova et al. 2007).

Chamois and Its Ecology

Origin, Systematic Classification, Mode of Life

The origin of chamois is rather old, deriving from an extinct antelope - *Pachygazela grangeri*, living in the Chinese Pliocene 12 million years ago. The oldest chamois, similar to contemporary species, occurred in Western Europe in the earliest Pleistocene. Well-founded findings are known from the late Interglaciale, a Neanderthal era, dating back to 115 thousand years B.P. (Blahout 1976, 1977). The animals similar in appearance to contemporary chamois were found in the younger Pleistocene –Würm glasier epoch (Blahout 1976). Based on palaeonthological findings, *Rupicapra* genus was for first time found in Europe in Central Pyrenees in midle Pleistocene (250 0000 – 150 0000 years ago) . The following Würm glatiation (80 000 – 12000 years ago) has not only seen an abundance of *Rupicapra pyrenaica*, but also *Rupicapra rupicapra*. Consequently, in the oldest Wüm (Masini and Lovari 1988), R. *pyrenaica* has split into two groups, i.e. R. *pyrenaica ornata* (Alpine Italian chamois) and R. *pyrenaica pyrenaica* (Spanish and French pyrenaica chamois). They are close subspecies lines, though differing morphologically, behaviourally and genetically from the Alpine chamois (Lovari a Scala 1980; Lovari 1981; Lovari a Locati 1991; Randi a Mucci 1997). Tatra chamois has diverged from the Alpine chamois approximately at the same level as Pyrenaica chamois and is very close to the Carpathian species in Romania. The Würm glatiation of the West Carpathians begun about 70 000 and lasted until 9 000 years B.P. This last glaciation is punctuated by a sequences of stadial (periods of colder temperatures) and interstadial (periods of warmer temperatures) events. It was probably during those warm periods of the Interstadial WI/II, in the tundra and steppe that the species *R. rupicapra* occurred in the Alps, Moravia, the Czech Republic, Slovakia and also in the south part of Poland (Musil 1985). Thus in the areas situated at lower altitudes the chamois could live in colder forests. At the time of the last oscilation, the Tatras country resembled a sparse taiga with forest-free areas. During this period (11 800 – 10 250 years B.P.) the Tatras vallyes were glaciated at 1300 – 1100 m above see level, but the peaks were not (Midriak 1983; Klimazsewski 1996). An important finding of fossil chamois from the Belianske Tatras dating back to 10 600 years ago indicates that the chamois could live in an Alpine altitudinal zone during this glaciation (Blahout 1977). Towards the end of the stadial WIII, the chamois lived both in the European mountains and on the planes in the north part of Italy, Slovenia, in Central part of Germany, the Czech Republic, Moravia, Slovakia and also in the Caucasus. With the commencement of the the Holocene (app. 10 000 B.P.) the climate of Tatras has swung towards warmer climatic conditions. During the Preboreal phase (10250 – 9100 B.P.), the Siberian taiga still existed in the Tatras with *Pinus sylvestris, Pinus cembra, Picea laris*. In the Boreal period (9100 – 7700 B.P.), the spruce stands were predominant, and other coniferous stands receded to extreme sites (Koštialik 1995; Klimazsewski 1996) The chamois had descended to lower altitudes, probably through avalanche gullies. In humid and relatively warmest period of the Holocene (7700 – 5100 B.P.) the upper tree line was 1800 m above sea level and the spruce stands prevailed in the Spiš region with the frequent peatbogs. The oak and fir forests were more frequent in the lower altitudes (Futák 1955). The chamois have adapted probably to the alpine zone as a results of the climatic changes in this period (Struhár 2001). Based on the well-known biology and food requirements of current chamois species, it can be said that the animals formed certain groups within the undisturbed environment of the Carpathians that lived in the forests while others resided on the Alpine meadows (Zelina 1956, Teren 1957).

Systematically, chamois belongs to the order *Arctiodyctyla* (even-toed ungulates), suborder *Ruminantia* (Ruminants), family *Bovidae* (Bovids), subfamily *Capra* (Goats), genus *Rupicapra* (Chamois) (Kratochvíl and Bartoš, 1954). Blahout (1977) reported three main chamois genera – the Alpine chamois, *Rupicapra rupicapra* (Linné, 1758), known as the original species in the Alps, the Pyrenees, Abruzzi, the Balkans, the Carpathians, Caucasia and the High Tatras, the Rocky mountain goat, *Oreamnos americanus,* (Blainville, 1816), endemic to the North America and the Alaska. The Himalayan goral, *Naemorhedus goral*, is common in Central Asia in the mountains on the border between China and Korea. According to the aforementioned author, besides three basic species, living in the separate mountain ranges, with no chance of cross-breeding, separate geographical breeds – or subspecies - have developed, differing in several anatomical features, such as the shape and size of the skull, horns' structure and colour of the coat. One of the Alpine subspecies is *Rupicapra rupicapra tatrica*, the Tatra chamois (Couturier (1938). On the other hand, Janiga and Zámečniková (2002) reffered to the fact that there are at least two known chamois species in Europe , thus denying the existence of only one species with ten geographical subspecies, as it is often stated today (Brtek 2001). *Rupicapra pyrenaica* (with subspecies *parva, pyrenaica* and *ornata*) from southwestern Europe is phylogeneticaly older species, while *Rupicapra rupicapra* (with subspecies *cartusiana, rupicapra, tatrica, carpatica, balcanica, asiatica* and *caucasica*) from northeastern Europe is younger (Grubb 1993).

Within the Slovak territory, the chamois is endemic solely to the High Tatras. Their nativeness to the High Tatras has been directly or indirectly proved on the basis of historical data reported by several authors (Sperfogel 1517; Zündt 1567; Buchholtz 1664; Towson 1797; Nowicki 1868 –ex Bohuš 1958). Couturier (1938) mentioned the occurernce and distribution of chamois in the former Czechoslovakia. He described their distribution as those living in the High Tatras and those occuring in the Alpi Zdiarske (Belianske Tatras) with the introduction of thier popular habitats. The skeleton of a subfosil chamois was found in the Belianske Tatras in 1974. It was the first finding that, besides other, confirms endemic occurrence of this subspecies in the High Tatras (Blahout 1977; Kováč 2002).

Taxonomic evaluation of the Tatra chamois, i.e. statistical evaluation of craniometric data, body weight and an extents, showed that Tatras chamois differ from the Carpathian and Alpine subspecies (Hanzák and Veselovský, 1960) and forms separate subspecies the Tatras chamois - Rupicapra rupicapra tatrica (Blahout, 1971). The individuals of the genus Rupicapra can be based on the methods of molecular biology, classified into two subspecies at least - R. rupicapra and R. pyrenaica. These analyses showed that Tatras chamois belong to genus R. rupicapra and represents an evolutionary distinct branch within this genus (Kmety et al. 2002; Rodriguez et al. 2009). The glacial relict and endemic subspecies Tatras chamois is strictly protected in the Tatra National Park, living in herds in the Alpine and subalpine zone, and preffering grassy downhills with scattered rocky parts. Herd size depends on the actual conditions and animal density. Each herd migrates within its territory according to certain rules (Blahout 1975, 1976; Chovancová 1985, 1990). There are three types of movements – daily, seasonal and irregular. Chamois alternate their stands only a little during the daily migration, they usually stay in same part of the valley, searching habitat with an optimal temperature and suitable grazing. Seasonal migration is connected with movement within broader area, though not exceeding the territory of one, or two neighbouring valleys. Irregular migration is triggered by extraordinary events, such as excessive mountain tourism or an enemy pursuing the individual chamois or the whole herd and forcing them to change habitats. Seasonal migratation of chamois to the higher altitudes of Alpine meadows in search of food occurs regulary during summer and autumn. Though they usually stay at the mountain ridges during winter, both Alpine and Tatra chamois often descend into lower altitudal zone if the conditions are extremely unfavourable. Other couses include favourable conditions in the spring, when chamois descend into lower spruce stands in search of food (Chovancová a Šoltésová 1988; Krokavec Jr.and Krokavec Sr. 1991; Babat and Ducaud 1997; Bögel et al. 1998). Chamois, however, need steep rocky formations for their safety (Chovancová 1985, 1990; Krause and Smidt 1997).

Ecological Relations of Chamois to Abiotic and Biotic Environment

The segmentation – configuration of the terrain, i. e. geomorphological and orographical conditions are rather important factors forming the relationship of chamois to abiotic environment. (Blahout 1976, 1977). The Belianske Tatras provide less suitable life conditions for chamois during summer due to little segmentation and steepness of the terrain, i.e. unsuficient protection against enemies, and an excessive turism. In winter, these conditions are more convenient (for peacefull condions, less snow and its faster thawing (Blahout 1976, 1977). When it comes to their protection, the terrain of the central part of the High Tatras is more suitable for the chamois. They prefer south-exposed wide valleys. Cold and deep side valleys with insufficient sunshine are less preferred. Some of the habitats preferred today are the same as those sought by chamois some 60 years ago (Bethlenfalvy 1937; Couturier 1938).

Climatic conditions also play an importan role. The behaviour of chamois when searching for suitable habitats is greatly influenced by the temperature of environment. The stands are changed with various altitudes and exposition. In the summer, the chamois climb up to highest parts of valleys, seeking northen, northwestern and northeastern sidehills. On the other hand, after long-lasting snowstorms and freezing animals climb down to lower altitudes to south, southwestern and southeastern downhills. These changes are regular and it is possible to predict the chamois movements (Blahout 1964, 1976, 1977; Chovancová 1985, 1990; Janiga &a Zámečníková 2002; Ksiažek 2002). Other important factors are searching for shade, cooling in the wet soil, flowing water, snow or agitated air. The chamois mostly search for places with the longest sunshine, such as glacier thresholds, moraines exposed to the south or southwest to avoid sharp wind. They search for the snow-free locations to rest and ruminate, and during harsh and severe conditions they find the shelter in caves, or overnahging rocks. Chamois prefer subalpine and alpine zone with unenclosed ground, allowing them to detect the danger in time. They are rather fearful in forest stands they stay there only for a short time, either during migration or while searching for food. Chamois try to avoid continuous zone of the scrub pine, they prefer zone of interspersed scrub pine for hiding their offspring and for the scent marking of the territory in the time of rut season.

Grassy and herbaceous vegetations covering mountain meadows are the food base for chamois. As the growing seasons in chamois biotopes is rather short, the animals have had to adapt to given conditions. In winters they feed on remains of grass and few species of perennial plants that they search for and dig from under the snow. Lichens present anothert important food component, having supposedly healing powers for chamois. The most intensive grazing occurrs in summer and animals will often choose the most tasty plant species, ingesting just the best and most tasteful parts (Blahout 1976; Chovancová and Šoltésová 1988). Tatra chamois can ingest toxic plants such as *Delphinium* spp. and *Veratrum lobelianum* with no physical harm. The later species, which grows in the Alps, Romanian Carpathians and Caucasus, is also very toxic for ruminants. Hadač (1960) commented on different properties of two forms of *Veratrum lobelianum*. The alkaloids of this plant can supposedly increase the digestion rate.

Typical animal species living together with chamois is Tatra marmot (*Marmota marmota latirostris*) (Kratochvíl, 1954). A relatively high degree of the coexistence has developed between these two species. Chamois are able to distinguish marmots' high pitched whistles, warning colony members of approaching enemy. The relationship between the chamois and deer and roe deer are indifferent. The beneficial coexistence exists between chamois and the songbirds, i.e. *Prunella modularis* and *Anthus spinoletta* (Blahout 1976; Chovancova 1995). The studies dealing with the relationship of the chamois and its predators are discussed in the following part.

Causes of Decrease in the Tatra Chamois Population

In the past, the chamois ranked amongst the species greatly attractive for hunters. As early as in the 15th century, a group of chamois hunters was formed in the SubTatras region –

poverty, starvation and alleged healing power of the bezoars from the chamois stomach had forced them to hunt the chamois, which has unfavourably resulted in their decreasing numbers. Poaching thrived in particular at the turn of the 19th century. Similarly to them, though for different resons, feudal sovereigns had taken to hunting chamois, competing for number of throphy animals. During his lifetime, Prince Augustus-Saxe-Cobourg (1845 – 1907), for instance, shot 3412 chamois. Juraj Bucholtz, the Slovak astronomer, studying the biology of chamois, reffered to the imbalance between the shooting and the population growth. The scientists such as Ludwik Zyszner, Eugeniusz Janota and Maksymilian Nowicki pointed out the critical status of the Tatras chamois population within 1851 – 1866 (Bohuš, 1966). In 1872, a law was passed on the Protection of chamois game in Hungary and in subsequent years several other conservation activities have contributed to their increasing numbers to 1000 animals in 1895. The World War I and World War II have decimated their numbers to 300 individuals. Declaration of the Tatras National Park (TANAP) in 1948 allowed a complex conservation of chamois, and thus the numbers of Tatras chamois increased 7 times within 15 years. Since the half of the 1960s, the numbers have decreased again to 160 individuals in 2000, raising a great concern about their future. In order to prevent the chamois extinction, during 1969 – 1976, there was an attempt to create viable population of the Tatra chamois outside their original area. The Tatra chamois has such been introduced to the central part of the Low Tatras National Park. Since then, due to comprehensive conservation measurements, the population has increased to 350 animals (Blahout, 1977; Kováč, 2002; Chovancová-personal communication). Chudík (1968a, b, c) was the first one to pay the attention to the mortality of chamois. He believed that the reasons for the population decline lay in the disrupted sex ratio as the result of the previous buck shooting. He stated that harsh climatic conditions during winter occurring several years in succession and increasing anthropic pressure have decimated the chamois population. Parasites also play an important role in reducing the population growth. Similar causes of the decline in the chamois population have been introduced by Blahout, 1975, 1977; Chovancová, 1985, 1990; Kováč 2002).

As for the biotic factors, Sokol (1970) considers the lynx to be the most important one to influence the decline of chamois. On the other hand, Betlenfalvy (1935) and Blahout (1976) believe the lynx to be an important selective-sanitary factor. The relationship between the lynx and free-living even-toed ungulates (deer, roe deer,chamois) was investigated by Bališ and Chudík (1970; 1976) who reported that the lynx accounted for app. 0.6% of the annual losses of Tatras chamois. Radúch (2002) found that not all individuals of current lynx population (30-35 animals), living in the territory of the TANAP, contribute to the predation pressure, just the animals entering the chamois biotopes in the subalpine altitudinal zone of the Tatras. Brtek (personnal communication) stated that the lynx is biologically progressive species specializing in predating smaller even-toed ungulates, including the chamois and therefore predating chamois is its natural feature, which is not in contradiction with its biological role as a predator in alpine ecosystems of the Slovak Carpathian. As the current chamois population within the TANAP is rather weakened, mainly due to an increasing human interference, the predating lynx can cause excessive losses of the chamois prey and decimate its population even further. In order to ease the predation pressure, certain number of individuals were shot in the important chamois biotopes.

The wolf plays lesser role in predating the chamois (Chudík 1968, 1969, 1974; Bališ 1969; Bališ and Chudík 1970; Sokol 1970; Blahout 1976). The harsh and rugged terrain of subalpine and alpine zone doesn't alow wolves to fully facilitate their vitality and mental abilities while using their hunting strategies. The wolf predominantly prey on the wild boar and the red deer and they represent 24 – 46 % of the total wolf prey (Brtek and Voskár 1987; Voskár 1993; Janiga and Hrkľová 2002; Strnádová 2002). Nadisan (1977) and Lovari (1977) also reported that a wolf seldom acts as a chamois predator. The fox, raven eagle and bear play little role in influencing the population numbers of chamois (Blahout 1976, 1977; Radúch, 2002).

Betlenfalvy (1935), Pavlovský (1935, 1936), Michelčík (1946), Zelina (1956), Bališ (1967), Blahout (1958), Chudík (1968) and Sokol (1968) studied the effect of abiotic factors on the chamois. Avalanches often contribute to mass mortality and physical injuries, along with falls from rock faces. Betlenfalvy (1935), Pavlovský (1935, 1936) and Venator already referred to negative anthropic impacts and uncontrolled turism that is very dangerous for the chamois. Blahout (1975, 1977) reported the problems concerning expansion of tourism in the TANAP and referred to negative impact of an anthropic pressure. The most common reaction to human presence in their territory is habitat change, which results in degradation of trophic factors, changes in escape distance, abandoning the usual rest sites. Intensive tourism also disturbs the territorial sovereignity due to mixing of the herds, thus creating social stress. In summer, the chamois spent 41 % of all their activities ruminating and resting. In winter, being forced to change their habitat, they often move to the locations with deep snow cover, weakening their physical constitution and abilities to withstand unfavourable and harsh weather conditions. All those effects are reflected into a daily rhytm change and two important peaks of daily activities overlap with peaks of morning and evening activities. Thus the usually diurnal animals become crepuscular ones. This can be a rather disturbing element, enhancing the negative impact of aforementioned factors (Blahout 1967; Janiga, 2002).

Chovancova (1985, 1990) and Gašinec (2002) analysed current negative anthropic impact and direct and indirect factors. Direct ones affect bionomy, and they divided them into two groups. The first group comprises factors affecting the animals from the earth surface, i.e. tourism, mountaineering, skialpinism, forest fruit picking and poaching. The second groupinclude factors affecting the animals from the air, i.e. aeroplanes, paragliding and rogallo flying. Hunting, environmental pollution and global climate changes comprise indirect factors. As a framework principles of protection against direct human activities they state securing the protection of chamois population by physical monitoring and guarding, securing the existence of their biotopes, regulation and control of tourism, sport activities, and by consitent observance of all provisions and acts on the nature and landscape protection.

Parasitic and Infectious Diseases

Parasitic and infectious diseases have contributed to chamois morbidity. In the following part the attention will be focused on the lung nematodes, gastrointestinal parasites and another pathogens of chamois in the different geographic areas with an emphasis on diseases of chamois in the Slovak national parks.

Lung Worms

Protostrongylids are parasites of the class Nematoda, order Strongylida, family Protostrongylidae. They are known as small lungworms because most of them parasitize the lungs of their hosts. Ruminants serve as their final hosts, along with leporids and very rarely carnivores. Protostrongylids have been known for a long time and studied since the beginning of the 19th, in 1802, when Frölich described a rabbit lung nematode, *Filaria pulmonalis* (Boev 1975). First reports on protostrongylids parasitizing wild ruminants date back to the end of the 19th century when enormous losses of chamois populations in Bavaria (area of lakes Schliersee and Tegernsee) and Tyrol (ducal Coburg hunting grounds, Pertisau) were related to protostrongylid lunworm infection (Mueller 1889; Gebauer 1932; Betlenfalvy 1937). Because of the location of many of the nematodes deep in the lung tissue, in bronchioles and alveoli, these helminthoses are difficult to cure. Protostrongylid parasitism causes decrease in amimal productivity (Thomson et al. 2000) and respiratory function (Berrag and Cabaret 1996; Jenkins et al. 2005), retardation in the development of young animals, delay in fur shedding, poor formation of the trophy of game (Mutafov et al. 1989) and mortality (Stroh 1936; Sattlerová-Štefančíková, 2005).

Papers dealing with chamois lung nematodes are scarce because material is very rarely available worldwide. The most frequent lunworms of chamois from various territories of Europe and New Zeland were species belonging to the genera *Muellerius*, *Neostrongylus* and *Protostrongylus*. *Dictyocaulus* spp. was less common and *Cystocaulus* spp. very rare (Table 1).

The data on lung nematodes of chamois had been incomplete and scarce in the territory of Slovakia for a long time. Mituch (1969), while studying helminth fauna of hoofed game in the TANAP, detected in chamois lung nematodes *Muellerius capillaris* and *Muellerius tenuispiculatus*, and within further studies Mituch (1974, 1989) also detected *Neostrongylus linearis*.

A considerably decreased number of Tatra chamois has evoked a whole range of activities aimed at its protection, including the study of parasites. The results of the research of ecology morphology, pathomorphology of lung nematodes, carried out during the past several years in the biotopes of Slovakia allowed to evaluate its effect on the chamois populations. Tatra chamois (*Rupicapra rupicapra tatrica*) living in the Tatras National Park (TANAP) and the Low Tatras National Park (NAPANT) are parasitized by the following nematodes species: *Muellerius capillaris*, *Neostrongylus linearis* and the specific chamois species *Muellerius tenuispiculatus*. In the Slovak Paradise National Park (NAPASR) and the Slovak National Park of Veľká Fatra (NAPAVF), in introduced chamois of an Alpine origin (*Rupicapra rupicapra rupicapra*), we found the same species of the lung nematodes, but in the chamois of Tatras subspecies and Alpine origin we also sporadically determined the geohelmith *Dictyocaulus viviparus* (Sattlerová-Štefančíková 1981, 1987; Štefančíková 1994, 1999, Štefančíková et al. 1999a, b, 2002; Sattlerová-Štefančíková 2005).

The morphological features of the distal ends of males are insufficient for the purpose of the differential diagnostics. The basic differential and diagnostic feature is spiculae, which in *M. capillaris* are rough, short and evenly long, in *M. tenuispiculatus* are thin, long and similar in length, in *N. linearis* the spiculae are thin, long but uneven in length. Larvae L1 allow - due to their unique morphology - the differential diagnosis, more common in routine practice. In both species of *Muellerius*, the tail has a blade extrusion, twisted around its own axis and thus the intens are formed on the distal end of the tail, in *M. tenuispiculatus* rather notable. In both species, a spiked extrusion is formed on the dorsal side, behind which is in *M. tenuispiculatus* deeper intent. The tail end of *N. linearis* is finished by dagger-shaped appendage with two transverse ligatures. The dorsal spine is poorly developed. (Fig. 2, 3, 4).

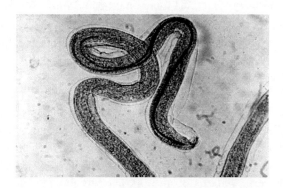

Figure 2. M. tenuispilucatus – male with thin and evenly long spiculae.

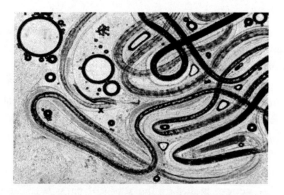

Figure 3. M. capillaris – male with rough, short and evenly long spiculae.

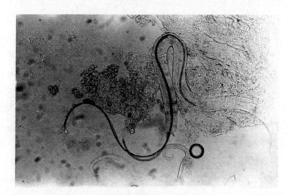

Figure 4. N. lineais – male with thin and unevenly long

By autopsies and larvoscopic examination of chamois faeces in TANAP in 1977 – 1980, 100 % prevalence of lung nematodes was determined. The number of L1 larvae per gram of faecal samples was permanently high in all seasons of the year. Prevailing species in the central granite complex of the High Tatras were *M. capillaris* and *M. tenuispiculatus*, and in the limestone part in Belianske Tatras *N. linearis* prevailed. The re-examination of the helminth status in 1997 shown the same species of lung nematodes but, the prevalence was substantially lower (48.4 %) with prevailing *Muellerius* spp. (45.6 %) in both complex of Tatras (High, Belianske Tatras). *N. linearis* occurred only in 11.9 %. Mean number of larvae per gram of faeces was low (Sattlerová-Štefančíková 1981, 2005). In 2006 -2008 the prevalence increased again on 98% in the central granite compex of the High Tatras and 100% in the limestone part in the Belianske Tatras. Mean larvae number was also increased over 100 larvae per gram.

This status of lung nematodes in chamois of the TANAP is determined by several factors. Harsh climatic conditions occurring over several years at the time of the birth and the presence of predators and increasing anthropic interference have decimated the chamois population. While in 1964 – 1967 chamois population reached 600, in 2000 it was only 160, at present, it is just about 350 animals. The strong contamination of environment (toxic metal substances) has led to the decline of biodiversity of trophic base of chamois that contributed to decrease in reproduction of chamois population, vitality and the resistance; and its impact on the gene poll is still unravelled. (Chovancová and Šoltésová,, 1988; Hell and Chovancová, 1995). The decreased resistance has resulted in the disturbance of a host-parasite balance as a consequence of previous strong infections, which contributed to the decrease in chamois population and in turn has resulted in the lover dissemination of the lungworm propagative stages into the environment. These factors have limited the infectivity of the intermediate and definitive hosts and hence the infection rate in 1997 was lower than those recorded 20 years ago (Sattlerová-Štefančíková, 1981, 1987, 2005). At present, increasing degree of infection is probably connected with global climate changes, which is reflected in the protraction of life time of larvae L1 into the enviroment and the accelerating of larvae in intermediate hosts-snails.

During 1981 – 1988 in the Low Tatras National Park, larval stages L1 *Muellerius* spp. (*M. capillaris, M. tenuispuculatus*) prevailed in faecal samples, *D. viviparus* occurred only sporadically. The results of autopsy revealed the prevalence of 71.4 %, larvoscopic examination of faecal samples - 89.6 %. Mean larvae L1 pre gram faeces in particular years varied between 28.2 ±11.9 to 61.6 ±35.9. Larvoscopic re-examination of faeces in 1998 showed the similar prevalence as in the previous period - 88.5 %. The degree of the infection was also similar, remaining at the medium level - ranging from 25.4 ±10.9 to 57.8 ±32.6 larvae per gram of faeces. Besides *Muellerius* spp., we determined the species *N. linearis*, which was not found in the previous period. It is possible that this species could be transfered here with the alpine chamois living in the Slovak Paradise and Veľká Fatra as a result of its migrations into adjacent zone of the ocurrence of Tatra chamois in the Low Tatra National Park. This species occurred less in comparison to *Muellerius* spp. (*Muellerius* spp. 67.3 %, *N. linearis* 32.7%) (Štefančíková 1994; Sattlerová-Štefančíková 2005). Co-temporary helminthostatus of lung nematodes is at the same level than that of the previous period.

Examination of faecal samples collected over 1984 – 1993 from Alpine chamois in the Slovak Paradise National Park revealed the prevalence of 57.7 % in chamois herds. *Muellerius* spp. were predominant (56.5 %), *N. linearis* occurred in 29.2 % and *D. viviparus* was prevalent in 2.5 %. The prevalence of lung nematodes in chamois herds fluctuated in different years from 41.4 to 69.4 % and in different seasons from 33.3 to 83.3 %. Mean larval count per gram faeces in individual years ranged between 73.7 ± 60.1 and 148.5 ± 58.0. Chamois herds were significantly less infected in summer seasons (P < 0.05) than in other seasons, except for the summer of 1986, 1988, 1989 and winter of 1989. A necropsy confirmed the presence of adult stages of all three nematodes in the lung parenchyma, but in bronchi and trachea no adults of *D. viviparus* were observed. Mean larval count per gram lung tissue was high (152 ± 20.9 to 270 ± 19) (Štefančíková, 1999; Sattlerová-Štefančíková, 2005). At present, a total prevalence of lung neamatodes was 98.6 % and mean larvae L1 per gram ranged from 179.3 ± 32.1 to 285.2 ± 38.5 in individual localities.

Examination of faecal samples collected over 1981 – 1987 and a biopsy of the lungs from Alpine chamois in the National Park of Veľká Fatra revealed the total prevalence of 78.4 % in chamois herds. *Muellerius* spp. were predominant (54.7 %), in comparison to *N. linearis* (26.7 %). The prevalence of lung nematodes in chamois herds fluctuated with years from 61.6 to 91.1 % and with seasons of the year from 60.0 to 93.7 %. Mean larval count per gram faeces in individual years show a low to moderate degree of infection and ranged between 7.2 ± 6.5 and 95.1 ± 62.8. The re-examination of the helminth status in 2005 revealed the same species of lung nematodes. The mean prevalence in chamois herds was lower than that of previous period (62.1 %). *Muellerius* spp. were predominant (42.2 %) in comparison to *N. linearis* (21.3 %). The degree of the infection was also similar, remaining at the low to medium level (7.1 ± 6.2 to 46.3 ± 11.6) (Sattlerová-Štefančíková 2005). The current situation is approximately at the same level.

The study of the resistance of chamois lung nematodes (L1 larvae) against various physical factors under laboratory conditions revealed the following:

Muellerius spp. L1 larvae isolated from the faeces started loosing their viability at the temperature of -12^0 C after four hours, *N. linearis* L1 larvae after 12 hours. After 48 hours no larvae of the nematodes studied survived. L1 larvae of both the nematodes remained viable in the faeces for four days and several specimens remained viable for 12 days. *Muellerius* spp. L1 larvae isolated from faeces survived the temperature of 50^0 C for 10 minutes, *N. linearis* L1 larvae 15 minutes. All larvae died after 30 minutes. L1 larvae of studied nematodes survived in faeces at the temperature of 50^0 C for two days, until day six when no L1 larvae remained viable. *Muellerius* spp. and *N. linearis* L1 larvae withstood single dehydratation (temperature 22-24^{0}C) for two days. After three days more than half of the larvae revived. On day five all the larvae died. Repeated alternation of dryness and humidity killed the L1 larvae of both nematode species in 36 hours. In dried faeces at the temperature of 22-24^{0}C they remained viable for five months and several specimens survived even for seven months. *Muellerius* spp. and *N. linearis* L1 larvae survived in tap water for 11 weeks. All *Muellerius* spp. L1 larvae died at the end of week 16, *N. linearis* L1 larvae in the middle of week 16. Under direct solar radiation (August, an atmospheric temperature of 35^{0}C, surrounding substrate temperature of 45^{0}C), *Muellerius* spp. and *N. linearis* L1 larvae managed to survive for an hour. In the natural High Tatras conditions, L1 larvae of the lung nematodes were recovered after a year in accumulated faeces on the sites, using by chamois as shelters during severe winter and storms (Chamois cave) (Sattlerová-Štefančíková 1982).

Infective L3 larvae *M. capillaris, M. teniuspiculatus* and *N. linearis* survived at the temperature of -12^{0}C without dying for two hours, all the L3 larvae of studied nematodes died after five hours. They withstood the temperature of 50^{0}C without damage for 30 minutes. After 90-minute exposition to such temperature all L3 larvae died. Dehydratation at the temperature of 22 – 24^{0}C was withstood without damage for two days. After four days all the

L3 larvae died. In the tap water they survived without damage for 21 days until none of them remained viable on day 32. The larvae survived direct solar radiation for an hour. The comparison of the differences in the resistance of both L1 and L3 larvae of lung nematodes showed that L1 larvae tolerated low temperatures better, but L3 larvae survived higher temperatures. The humidity was tolerated better by L1 larvae. Both L1 and L3 larvae survived solar radiation and drying for approximately same time. The results mentioned above showed that ontogenetic stages of lung nematodes of chamois are adapted to successfully complete their life cycle in the extreme climatic conditions (Sattlerová-Štefančíková 2005).

The biotopes of the High Tatras differ from those of the Belianske Tatras, having a significant effect on the structure and density of the mollusc fauna. The granite bottom of the High Tatras with a continuous belt of spruce forests, producing acid substrate, do not provide a favourable living conditions for snails. On the other hand, in the limestone part of the Belianske Tatras, the snails are abundant. Their abundance, and the species diversity is decreasing with the higher altitude and in the Alpine zone (app. 1800 m above sea level) the snails are scarcely found on the granite. On the limestone – the species diversity is quite rich with a strong snail population. The results of the malacological excursions at the chamois sites proved that the snail populations in the Belianske Tatras are more diverse than those in the central complex of the High Tatras. The snails were collected from 1977 to 1980 from June till the half of August. The found species were as follows: *Clausilia grimmeri* (A. Schmidt, 1857), listed by Ložek (1956) as syn. *Clausilia dubia* var. *carpathica, Ena montana* (Draparnaud, 1801), *Cochlodina laminata* (Montagu, 1803), *Collumella edentula* (Draparnaud, 1801) and *Pyramidula rupestris* (Draparnaud, 1801). These species were mostly found at 1500-1800 m above sea level (subalpine zone) under small rocks or on the ground parts of the plants. Another species was dominant *Helicigona faustina* (Rossmäsler, 1835). They were found high in the Alpine zone over 2000 m above sea level, and on the tops of the mountain peaks – either hiding in the rocks screes or basking in the sun. According to our experience, the snail numbers in the Belianske Tatras during our study were not decreasing with the growing altitude. The areas covered with parched and decomposed plants provide due to their air-thermic isolation an ideal place for snail hibernation. The species *Halicigona faustina* occurred in these areas even when they were covered by snow - during collections in June, when at the lower locations of chamois pastures new vegetation already emerged, while the higher locations were still under snow. The chamois become infected in the High Tatras during seasonal migrations, in particular in spring, when they are going down to the lover altitudes in search of food and at irregular migrations during the tourist season. These facts are supported by the determined prevalence of lung nematodes in chamois herds, which varied in the different localities of this part of the Tatras. On the other hand, in the Belianske Tatras, the prevalence in different localities during 1977 - 1980 and in 1997 did not varied as much, as the infections can occur throughout the whole year. Experimental infection of various intermediate host species with individual lung nematodes showed that the larvae reached the infective stage in all snails collected on the chamois sites (*Muellerius* spp. on day 35 and *N. linearis* on day 21).

Despite the geomorphologic and climatic similarity of the chamois biotopes in both the High and Low Tatras, the current area of chamois distribution in Low Tatras does not contain calciphile plants since this is an area without the limestone-dolomite substrate. This affects both the chamois trophic base and the occurrence of snails – the intermediate hosts of lungworms. They are abundant mainly in lower zones, but towards the alpine zone the number of snails is rapidlly decreasing and over 1500 m above sea level they occur only sporadically. The moderate degree of infection in chamois in this reservation showed that animals are predominantly infected during their seasonal migrations, when they go lower - below the upper forest line, mainly in spring, but also throughout the whole during their irregular migrations.

Rather rugged terrain of the Slovak Paradise, temperature and humidity differences between cold and humid canyons and sun–exposed rocks with a warm and nutrient rich limestone base determine the abundance and species diversity of plant cover, which is also reflected in the number and species composition of malacofauna. This territory yielded 114 snail species (Šteffek 1975). The most abundant species, besides many others, in this territory are: *Helicigona faustina* (Rossmasler, 1835), *Clausilia dubia* (Draparnaud, 1805), *Cochlodina laminata* (Montagu, 1803), *Succinea putris* (Linné, 1758), *Ena monntana* (Draparnaud, 1801) (Šofránková 1982). These species also occur in numerous populations in the area of the High Tatras and they have been experimentally proved to play the role of intermediate hosts. Species *H. faustina* contained lung nematode larvae at different stage of development also under natural conditions. Similarly, in the National Park of Veľká Fatra dolomitic and limestone substrates create appropriate living conditions for the snails and thus creating possibility for chamois to become infected with lung nematodes.

The variety of snail species in the all of the Slovak National Parks and their great ability to serve as the intermediate hosts of lung nematodes in different biotopes create the appropriate conditions for the development of lung nematodes. Long lifetime and great reproduction abilities of parasites in the lung, great resistance of L1 larvae in the environment and opportunity to hibernate in snails, enables them to perservere in all chamois biotopes in Slovakia.

Based on the pathomorphological changes in chamois suffering from lung nematodoses, the changes can be divided macroscopically into four morphologically distinct groups:

Petechia – petechial haemorhage, occurring after an active migration of postinfection larvae from the blood stream to the lung tissue. Airless areas, parasitic lung changes occurring during the parasite transition into mature and reproductive stage. They are of various size, changing shape and colour, 1-3 mm long, of reddish brown colour or grey and in case they are located under thickening pleura, they are a murky milky white shade and opalescing. The shape is irregular, often lobed. Productive nodules are the most common changes of the lung tissues. They occur on the surface of the lungs as irregular , diamond shaped, less rectangular and scarcely lobbed nodules , length 1-4 cm, moderately protruding above the surface. Their patchy pinkish and grey or brown and grey colour is a distinctive feature. Emphysematous areas prevailed over airless. Indurated foci were noted only on the apical ends of diaphramatic lung lobes as slightly protruding foci with homogenous and solid elastic consistence, of reddish brown colour, in fibrosis of pleura of dirty brown colour, length from 2 to 5 cm and with irregular lobbed margin, penetrating deeply into the lung parenchyma, and often affecting whole apical end (Sattlerová-Štefančíková 1982, 1987, 2005; Švarc 1982, 1984).

Gastrointestinal Parasites

Similarly to lung nematodes, gastrointestinal parasites can also affect adversely the chamois health. Together with mentioned climatic factors, which affect the development of parasite propagative stages in environment, they degrade hosts' health reflecting on disturbance of a host-parasite balance. Thus the parasites can cause the losses in chamois population, mainly due to coccidiosis in young animals.

Gastrointestinal parasites were studied by many authors in the different European territories (Tab.2a,b) Gebauer (1932) found eight parasitic species in Alpine chamois from Austria, but Kutzer and Hinaidy (1969) detected as many as 22 species within the same territory, including *Ostertagia leptospicularis, Skrijabinagia* spp., which were found solely by aforementioned authors. In France, in two geographic subspecies (*Rupicapra rupicapra, R. rupicapra pyrenaica*) the detected fauna of gastrointestinal worms was almost identical, considering the number of species, but differed qualitatively. *Rupicapra pyrenaica* harboured larval stages of *Echinoccocus granulosus* (Donat et al. 1989), though they were not recorded in the Alpine subspecies (Hugonnet and Euzéby 1980). This cestode was also found in chamois from former Yugoslavia (Delic and Cancovic 1961), in the Carpathian subspecies in Romania (Almasan and Nestorov, 1972) and in the Alpine subspecies in the Czech Republic and Moravia (Kotrlý and Kotrlá 1977). *Dicrocoelium dentriticum*, a rather rare chamois parasite, was detected besides France (Hugonnet and Euzéby 1980), in Austria (Kutzer and Hinaidy 1969) and Switzerland (Salzman and Hörning 1974). *Fasciola hepatica* has rarely been found in chamois (Gebauer 1932; Kutzer and Hinaidy 1969; Siko and Negus 1988). *Oesophagostomum radiatum* and *Bunostomum trigonocephalum* were found only in the Alpine subspecies in Italy (Genchi et al. 1982, 1984, 1985). Other gastrointestinal helminths mentioned in Table occurred more often in chamois of the mentioned geographic races.

Gastrointestinal helminths in chamois of the High Tatras were studied in the framework of study of helminth fauna of hoofed ruminants by Mituch (1969; 1974), Mituch et al. 1984, 1985, 1989). In the former study, Mituch (1969) reported eight helminth species, with predominant cestode *Cysticercus tenuicollis* (30 %) and nematode Nematodirus helvetianus (22 %). He also reported trematode *Fasciola hepatica* (7.6 %). The following study (Mituch 1974) comprised almost identical species proportion, though with prevailing *Trichostrongylus colubriformis* and *Nematodirus helvetianus* instead of *Nematodirus filicollis*. *Ostertagia ostertagi* was identified only in a single case (Mituch et al. 1985). The following studies (Mituch et al. 1984, 1985, 1989) also reported *Chabertia ovina* and *Haemonchus contortus*. Špeník (1975) found only five species with predominant coccidia *Eimeria* spp. (40 %) and rather high prevalence of *Ostertagia circumcinta* (25 %). Gastrointestinal helminth fauna of the Alpine chamois within the territory of the Slovak Paradise National Park is almost identical to those of the Tatra Chamois (Krokavec Jr. and Krokavec Sr. 1991, Ciberej et al. 1997). Unlike other authors, who studied this problem in the Tatra Chamois, Krokavec Jr. and Krokavec Sr. (l.c.) found, besides others, *Dicrocoelium dentriticum* (7,4 %) in the Alpine chamois (Tab.3).

Later, gastrointestinal nematodes in chamois and deer and coccidia in chamois were studied by Krupicer et al. 1998, 2004; Štefančíková et al., 2002; Sattlerová-Štefančíková, 2005).

Table 1. Survey of occurrence of long nematodes in chamois of different geographical races and areas by authors (modified Diez et al. 1990)

Country	Reservation	Authors	1	2	Dyctiocaulidae 3	%	3	Protostrongylidae 4	5	6	7
Austria	Alp	Gebauer (1932)	*	L		*	+		+	+	+
	Alp	Kutzer and Hinaidy (1969)	*	L	0				+	+	+
	Alp	Schröder 1971 (cit. Salzman and Hörning 1974)	10	L	0	35			+	+	+
Czech, Slovakia	Jeseniky, Česká Kamenica, High Tatras	Kotrlý (1967)	*	L	0						
	Jeseniky, Česká Kamenica	Kotrlý (1958)	*	L	0	39,5			+	+	+
	Jeseniky, H. Tatras	Erhardová (1957)	*	L	*	*	+		+	+	+
	Jeseniky, Česká Kamenica, Hig Tatras	Kotrlý (1964)	*	L	0	36,2			+	+	+
	High Tatras, Jeseniky	Erhardová and Ryšavý (1967)	*	*	*	*					
	Jeseniky, Česká Kamenica, High Tatras	Kotrlý (1970)	*	L	0	*	+		+	+	+
Czech, Slovakia	Jeseniky	CHroust (1991)	63	L/F	3,4–8,3	41,2/52,4	+		+	+	+
France	Bauges	Hugonnet and Euzéby (1980)	*	L	*	*	+		+	+	+
	Vanoise	Hugonnet et al. (1981)	*	L	0	*	+		+		+
	Bauges	Nocture (1986)	51	L/F	0	35,2			+	+	+
	Roc Blanc-Carlit massif	Donat et al. (1989)	59	L/F	0	*			+	+	+
Georgia	Caucasus	Rodonová (1962)	*	L	0	*					
	Caucasus	Gurčiani (1967)	*	L	0	*	+		+		+
Germany	Alp	Knaus and Schröder (1960)	*	*	*	*	+		+	+	+
	Alp	Salzmann and Hörning (1974)	58	L/F	15,2	33,6/8,3	+		+	+	+
	Alp	Stroh (1936)	100	L	0	98			+	+	+
Italy	Gran Paradiso	Balbo et al. (1975)	71	L	0	8,4			+	+	+
	Abruzzo	Cancrini et al. (1985)	21	L	0	47,6			+	+	+
	Val Belviso	Genchi et al. (1984)	20	L	0,8	95	+	+	+	+	+
	Stelvio		18	L	0	88,8		+	+	+	+
New Zeland		Clark and Clarke (1981)	28	L	*	*	+		+		+
Romania	Carpathians	Almasan and Nesterov (1972)	*	*	*	*	+				+
		Siko and Negus (1988)	3600	F	5	*	+		+	+	+
Russia	central Caucasus	Pupkov (1971)	*	L	0	*					+
	Caucasus	Ruchljadev (1950)	*	L	0	*			+	+	+
Slovenia		Bidoveč et al. (1985)	1232	L	0,4	83,2	+		+	+	+
		Brglez et al. (cit. Hugonnet and Euzéby (1980)	*	L	0	*			+	+	+
	Treskavica	Delic and Canković (1961)	2	L	0	*			+	+	+
Scotland		Dunn (1969)	*	*	*	*	+		+	+	+
Spanish	Mampodre, Riano	Diez et al. (1984)	14	L	0	100	+	+	+	+	+
	Reres	Diez et al. (1987)	15	L/F	0	100	+		+	+	+
		Diez et al. (1990)	66	L/F	0	90,5/93,5	+		+	+	+
Switzerland	Alp	Dollinger (1974)	213	L	0,5	67,7	+		+	+	+
	Alp	Hörning (1975)	*	*	*	*	+		+	+	+
Jura		Hörning and Wandeler (1968)	101	L	3,9	59	+		+	+	+
		Salzmann and Hörning (1974)	75	L			+	*			+

1 No. examined chamois; Examined material (L = lung, F = faeces); 3 Dictyocaulus spp.; 4 Cystocaulus spp.; 5 Muellerius spp.; 6 Neostrongylus linearis; 7 Protostrongylus spp.; +closely non specified species *Authors do not specify number studied chamois, material and prevalence

Table 2a. Survey of occurrence of gastrointestinal parasites of chamois of different races and areas by authors.

Country: Reservation: Authors Material:	Austria Alp Gebauer (1932) Autopsy	Austria Kutzer and Hinaidy (1969) Autopsy	Switzerland Alp Salzman and Hörning (1974) Autopsy	France Bauges Hugonnet and Euzéby (1980) Autopsy/Feaces	France Pyrenees Donat et al. (1989) Feaces	Italy Alp Genchi et al. (1984, 1985) Autopsy
Helminth :						
Moniezia benedeni	—	+	—			
Moniezia expansa	—	+	—	—	—	—
Avittelina centripunctata	—	+	—	—	—	—
Cysticercus tenuicollis	+	+	—	+	—	—
Echinococcus unilocularis	—	—	2,20%	—	+	—
Fasciola hepatica	+	+	—	—	—	—
Dicrocoelium lanceolatum	—	+	+	+	—	—
Bunostomum trigonocephalum	—	—	—	—	—	97%
Capillaria bovis	—	+	2,20%	—	—	—
Cooperia pectinata	—	—	—	—	—	—
Hemonchus concortus	+	+	24,10%	+	+	17%
Nematodirus filicollis	—	+	15.5 %	+	—	—
Nematodirus helvetianus	—	—	—	—	—	19%
Nematodirus rupicapra	—	—	—	—	—	11%
Nematodirus spathiger	—	—	—	—	—	8%
Nematodirus sp.	—	—	—	—	42,60%	—
Marshallagia marshalli	—	+	—	+	—	—
Ostertagia circumcinta	+	+	—	+	—	—
Ostertagia ostertagi	+	+	4,30%	+	—	28%
Ostertagia occidentalis	—	+	—	—	—	58%
Ostertagia leptospicularis	—	+	—	—	—	—
Ostertagia spp.	—	—	—	—	46%	—
Skrjabinagia lyrata	—	+	—	—	—	—
Skrjabinagia kochlida	—	+	—	—	—	—
Spiculopteragia spiculoptera	—	—	—	—	—	—
Skrjabinema sp.	—	+	—	+	—	11%
Trichuris ovis	—	+	23,90%	+	—	25%
Trichuris globulosa	—	+	—	—	—	17%
Trichostrongylus vitrinus	+	+	—	+	+	14%
Trichostrongylus axei	+	+	—	+	—	17%
Trichostrongylus colubriformis	—	—	8,70%	—	—	14%
Eimeria spp.	+	+	27,30%	—	—	—

Table 2b. Survey of occurrence of gastrointestinal parasites of chamois of different races and areas by authors

Country:	Romania	Romania	Ygoslavia	Ygoslavia	Czech, Slovakia	Czech
Reservation:	Carpathians	Carpathians	Trescavica	Beograd (ZOO)	Lužické hory, Jeseníky,	Jeseníky
Authors	Almasan and Nestorov (1972)	Siko and Negus (1988)	Delic and Cankovic (1961)	Nesic et al. (1992)	H. Tatras	Chroust (1991)
Material:	Autopsy /Feaces	Feaces	Autopsy/ Feaces	Feaces	Autopsy/ Feaces	Autopsy/ Feaces
Helminth :						
Moniezia benedeni	+	+	+	—	+	—
Moniezia expansa	—	—	+	—	—	—
Avittelina centripunctata	—	—	+	—	—	—
Cysticercus tenuicollis	+	—	+	—	—	—
Echinococcus unilocularis	+	—	+	—	+	—
Fasciola hepatica	—	+	—	—	—	—
Dicrocoelium lanceolatum	—	—	—	—	—	—
Bunostomum trigonocephalum	—	—	—	—	—	—
Capillaria bovis	—	—	—	—	—	2%
Cooperia pectinata	—	—	—	—	—	11,50%
Hemonchus concortus	+	+	+	+	+	26,30%
Nematodirus filicollis	—	—	—	+	—	42,50%
Nematodirus helvetianus	—	—	—	—	—	—
Nematodirus rupicapra	—	—	—	—	—	—
Nematodirus spathiger	—	—	—	—	—	—
Nematodirus sp.	—	—	—	—	—	—
Marshallagia marshalli	—	—	—	—	—	—
Ostertagia circumcinta	+	+	+	—	+	12%
Ostertagia ostertagi	+	—	—	—	+	39,80%
Ostertagia occidentalis	—	—	—	—	—	—
Ostertagia leptospicularis	—	—	—	—	—	52,60%
Ostertagia spp.	—	—	—	—	—	—
Skrjabinagia lyrata	—	—	—	—	—	—
Skrjabinagia kochlida	—	—	—	—	—	—
Spiculopteragia spiculoptera	—	—	—	—	—	31,70%
Skrjabinema sp.	—	—	—	—	—	18,60%
Trichuris ovis	—	—	—	—	—	12,50%
Trichuris globulosa	—	—	—	—	—	—
Trichostrongylus vitrinus	+	+	+	+	+	24,80%
Trichostrongylus axei	—	+	—	—	—	—
Trichostrongylus colubriformis	—	—	+	—	+	25,50%
Eimeria spp.	+	+	+	—	+	37,90%

Table 3. Survey of gastrointesinal parasites of chamois in Slovakia by authors

Reservation:	TANAP	TANAP	TANAP	TANAP	TANAP	TANAP	TANAP	NAPASR	NAPASR
Authors	Mituch (1969)	Mituch (1974)	Mituch et al. 1984	Mituch et al. 1985	Mituch et al. 1989	Špenik (1975)	Rajský and Beladičová (1987)	Krokavec Jr. and Krokavec sr. (1991)	Ciberej et al. 1997
Material:	Autopsy	Autopsy	Autopsy	Autopsy	Autopsy/Faeces	Autopsy/Faeces	Faeces	Autopsy/Faeces	Autopsy/Faeces
Helminth									
Cysticercus tenuicolis	30%	+	—	—	35%	—	—	—	—
Moniezia benedeni	15%	+	+	+	+	—	11.60%	—	—
Fasciola hepatica	7.60%	+	—	—	—	—	—	—	—
Dicrocoelium dentriticum	—	—	—	—	—	—	—	7.40%	—
Hemonchus concortus	—	—	+	+	10%	—	11.60%	16.60%	—
Chabertia ovina	—	—	+	+	25%	—	55.80%	24.10%	18.20%
Nematodirus filicollis	7.60%	+	+	+	5%	5%	—	—	—
Nematodirus spatiger	—	—	+	+	5%	—	—	—	—
Nematodirus helvetianus	22%	—	+	—	+	—	—	—	—
Nematodirus sp.	—	—	—	+	—	—	18.60%	14.80%	22.70%
Oastertagia sp.	—	—	—	—	—	—	16.20%	18.50%	—
Ostertagia circumcinta	15%	+	+	+	—	25%	—	—	—
Ostertagia ostertagi	—	—	—	+	—	—	—	—	—
Oesofagostomum venulosum	15%	+	+	—	10%	5%	13.90%	25.90%	9.10%
Oesofagostomum sp.	—	—	—	+	—	—	—	—	—
Trichuris ovis	7.6	+	+	—	—	15%	6.90%	16.60%	18.20%
Trichostrongylus colubriformis	—	+	+	—	5%	—	—	—	50%
Trichostrongylus axei	—	—	—	—	5%	—	—	—	—
Spiculopteragia spiculoptera	—	—	+	+	—	—	—	—	—
Trichostrongylus sp.	—	+	+	+	—	—	11.60%	9.20%	—
Strongylus sp.	—	—	+	+	—	—	—	—	—
Bunostomum trigonocephalum	—	—	—	—	5%	—	—	—	—
Bunostomum sp.	—	—	+	+	—	—	—	—	—
Eimeria spp.	—	—	—	—	—	40%	—	11.10%	—

The study of gastrointestinal nematodes of chamois and contact animals - deer - in the TANAP showed that both ruminant species harboured the same genera of gastrointestinal nematodes. In deer, the dominant genera were *Chabertia* spp. (21.8%) and *Trichostrongylus* spp. (19.5%). The biotopes in the High and Belianske Tatras showed the similar trend. In chamois generally, the dominant genera – similarly as in deer, were *Trichostrongylus* spp. (12.6%), *Chabertia* spp. (10.3%) their prevalence being significantly lower than in deer. In the High Tatras, *Chabertia* and *Trichsotrongylus* (11.2%) dominated and in the Belianske Tatras it was *Trichostrongylus* (16.9%) and *Oesophagostomum* (10.8%). The total prevalence of gastrointestinal helminths was lower in chamois (45.2%) than in deer (73.6%) Coccidia oocysts were found in chamois in the High Tatras in 5.3 %. In NAPANT, the gastrointestinal nematodes were represented by the dominant *Chabertia* spp. (73.0%) and *Trichostrongylus* spp. (53.8%). Cestodes were rare in the chamois from NAPANT (*Monezia* spp. 7.6%), though the prevalence of *Eimeria* spp. was high (80.7%). For the first time in the ecological conditions of Slovakia 4 *Eimeria* species have been determined there: *E. rupicaprae, E.riedmuulleri, E.yakimoffmatschoulski* and *E. alpine*. In the NAPASR, the highest prevalence was observed in the genera *Oesophagostomum* (52.6%), *Trichostrongylus* (63.8%) and *Eimeria* (60%). Gastrointestinal and protozoal parasitoses of chamois population in the Slovak biotopes are negatively affected by climatic and anthropic factors.

Bacterial Pathogens

For the first time within the Slovak territory, *Anaplasma phagocythoiphilum* was found in Alpine chamois from the Slovak Paradise (Vichová et al., 2008). In Switzerland, bacterium *Mycoplasma conjunctivae* caused outbreaks in Alpine chamois within an area of 1590 km^2 from 550 to 3200 m a.s.l. Many aninimals had to be put down due to their subsequent blindness and deteriorated health. An estimated mortality was app. 25 % (Citterio et al., 2003). For the first time within the same territory, *Babesia divergens/Babesia capreoli* have been detected in a chamois corpse (Hoby, 2007). *Salmonella enterica* caused the death of the chamois from the Tyrol Alps (Glawischnig et al., 2000). In Italy, the chamois died as a result of *Actinomyces* infection (Radaelli et al., 2007). Pioz et al. (2008) studied the influence of the weather and the bacterial pathogens (*Salmonella enterica, Chlamidophila abortus, Cocxiella burnetii*) on the reproduction of chamois. He identified the confounding effect of weather and parasitism on fecundity in a natural population and he concluded that after accounting for density, the prevalence of antibodies against the three bacteria explained 36% of the annual variation in reproductive success, and weather conditions explained an additional 31%.

Conclusion

Reducing the biodiversity of ecosystems is currently of great global concern that has been expressed in many international documents, programmes and strategies, such as the **Convention on Biological Diversity**, known informally as the **Biodiversity Convention**, adopted at the The United Nations *Conference* on Environment and Development in Rio de Janeiro in June 1992. Those and many other documents and treaties cover a vast array of problems, attempting to convey the message to man about necessity to protect biodiversity and biological resources and reffering to human moral standards, respect for all life and responsibility towards future generations. As the Tatra chamois population has regularly fluctuated, it requires systematical and complex monitoring and observation. The protection of genes and eliminating the negative human impact on their biotope are considered the substantial part of the problems relating to the conservation of this rare subspecies.

This work was supported by the National Environmental Fund and the grant agency for science VEGA grant No 2/0042/08.

References

Aulagnier, S., Giannatos, G., & Herrero, J. (2008). *Rupicapra rupicapra*. In: IUCN 2008. 2008 IUCN Red List of Threatened Species. <www.iucnredlist.org>. Downloaded on 05 March 2009.

Almasan, H., & Nestorov, V. (1972). Beitrag zur Kenntnis der Parasiten-Fauna der Gemse (*Rupicapra rupicapra carpatica* Couturier (1938) in den rumänischen Karpaten. *Z. Jagdwiss.*, **18**, 103-106.

Babat, G., & Ducaud, O. Impact of ungulates on fir natural regeneration of mixed mountain forests in a Northern French Alp area: Role of ecological conditions. *Volume of Abstracts of the 2nd World Conference on Mountain Ungulates*, 5-7th May 1977, 1997 s. 13.-14. Saint Vincent, Aosta.

Balbo, T. (1973). Indagini sulla situazione parassitologica nei mammiferi del Parco Nazionale del Gran Paradiso. *Parasitologia*, **15**, 301 – 312.

Balbo, T., Costantini, R., & Peracino, V. (1975). Indagini sulla diffusione dei nematodi polmonari nello stambecco (*Capra ibex*) e nel camoscio (*Rupicapra rupicapra*) del Parco Nationale del gran Paradiso e della Riserva di Valdieri. *Parasitologia*, **17**, 65 – 68.

Bališ, M. Animal kingdom of Tatra National Park. In Conception of Tatra National Park (ed. Pacanovský M.), *Obzor Bratislava*, 1967, pp.73-113.

Bališ, M., & Chudík, I. (1976). The Eurasian lynx (*Lynx lynx* L) share on losses in free-living ungulates in the TANAP. *Zborník prác o Tatranskom národnom parku*, **18**, 66-79. (In Slovak)

Berrag, B., Cabaret, J. (1996). Impaired pulmonary gas Exchange in ewes naturally infected by small lunworms.. *J. Parasitol.*, **26**, 1397-1400.

Bethelnfalvy, E. Tatra chamois: Naša zverina. Academia Bratislava, 1935, p. 227 – 232. (In Slovak)

Bethelnfalvy, E. Die Tierwelt der Hohen Tatra. Spišské Podhradie, 1937, p. 9-22.

Bidoveč, A., Valentinič, S., & Kuses, M. Parasitic pneumonia in chamois (*Rupicapra rupicapra* L) in Slovenia pp. 240-242. In *Lovary S. (ed) The Biology and management of mountain ungulates*. Lomdon, 1985, Croom Helm.

Blahout, M. (1958). Research on chamois ecology in the Podbanské nature reserve in 1956 – 57. *Zborník prác o TANAP*, **2**, 134-174, Osveta Martin. (In Slovak)

Blahout, M. Ecology of chamois. *Final project report*, 1965, Tatranská Lomnica. (In Slovak)

Blahout, M. (1969). Contribution to the chamois ecology in the TANAP. *Zborník prác o TANAP*, **11**, 381-389, Osveta Martin. (In Slovak)

Blahout, M. (1969). The impact of anthropic activities on life of some animal species in the TANAP. *Čsl. ochrana prírody*, **8**, 313-323 Príroda Bratislava. (In Slovak)

Blahout, M. (1972). Current status and prospects for development of chamois breeding in the Tatras National Park. *Lesnícky časopis*, **18**, 401-407.(In Slovak)

Blahout, M. Ecology of the Chamois. *Final report. The TANAP Research Centre*, 1975, Tatranská Lomnica.(In Slovak)

Blahout, M. Causes of decrease in chamois population in the TANAP. *Final report of the state research plan,* 1976, Výskumná stanica a múzeum Tatranského národného parku. Tatranská Lomnica.(In Slovak)

Blahout, M. *Chamois game. Príroda Bratislava*, 1976, p.171. (In Slovak)

Blahout, M. (1977). Some anthropic effects on chamois in the Tatras National Park. *Zborník prác o TANAP,* **19**, 127-168, Osveta Martin. (In Slovak)

Boev, S.N.: Legočnyje namatody kopytnych životnych Kazachstana. Izd-vo AN Kaz SSR, Alma-Ata, 1957, 177.

Bohuš, I. Contribution to the history of chamois and its conservation in the High Tatras. *Tatranské noviny*, 1956, 31,3, 32,3, 33, 3, 36, 3-4, 37, 3. (In Slovak)

Bohuš, I. (1958). Contribution to the history of Ibex in the Tatras. *Zborník prác o TANAP*, **2**, 148-159. Osveta Martin. (In Slovak)

Bohuš, I. (1966). The First in the High Tatras. *Krásy Slovenska*, **43**, 260 - 262. (In Slovak)

Bögel, R., Frühwald, B., Lotz, A., & Walzer, C. Habitat use and population management of chamois *Rupicapra rupicapra* in Berchtesgaden National Park 2nd World Conference on Mountain Ungulates, 1997, 13-22.

Brtek, Ľ. (1983). Problems with the Chamois. *Vysoké Tatry*, **22**, 10 - 11. (In Slovak)

Brtek, Ľ. (2001). Some controversial questions concerning rare and exploited animal species. *Správy Slovenskej zoologickej spoločnosti*, **19**, 33-36. (In Slovak)

Brtek ,Ľ., & Voskár, J. (1987). Food biology of wolf in the Slovak Carpathians. *Biológia,* **10**, 985-990. (In Slovak)

Cancrini, G., Iori, A., Rossi, L., & Fico, R.: Occurence of pulmonari and gastrointestinal nematodes in the Abruzzo chamois In: *The biology and management of mountain ungulates. S. Lovary (edit.), Croom Helm, London,* 1985, 256 - 257

CAPRINAE SPECIALIST GROUP 2000. *Rupicapra rupicapra* ssp. *tatrica*. In: IUCN 2006. *2006 IUCN Red List of Threatened Species.* <www.iucnredlist.org>. Download on 21 April 2007.

Ciberej, J., Letková, V., & Kačúr, M. (1997). Helminth fauna of alpine *chamois (Rupicapra rupicapra rupicapra*) in area Paradise. *Slov. vet. čas.*, **22**, 301 – 302

Citterio, C. V., Luzzago, C., Sala, M., Sironi, G., Gatti, P., Gaffuri, A., & Lanfranchi, P. (2003). Serological study of a population of alpine chamois (*Rupicapra r. rupicapra*) affected by an outbreak of respiratory disease. *The Veterinary Record*, **153**, 592–596.

Clark, W.C., & Clarke C.M.H. (1981). Parasites of chamois in New Zeland. *New Zeland Vet. J.* **29**, 144.

Couturie,r M. A. J.: *Le Chamois*, 1938. Grenoble (France): B. Arthaud-Editeur.

Délic, S., & Cankovič M. (1961). Prilog poznavanju parasitofaune divikoze (*Rupicapra rupicapra* L.) sa producja planine Treskavice. *Veterinaria Sarajevo*, **10**, 483-485.

Diez, P., Diez, N., & Anton, A. Aportaciones al conocimiento de los nematodos pulmonares del rebeco (*Rupicapra rupicapra*) de la Cordillera Cantábrica. *IV Reunion annal de la asociation da Parasitologos Españoles*, 26 - 28 septiembre, **61**, 1984, Madrid

Diez, P., Diez, N., Anton, A., & Morrondo, M.P. Principales problemas parasitarios del rebeco en la Cordillera Cantábrica. *Actas de las Jornadas de Estudio sobre la Montaňa.* Edit. URZ, 1987, 337 – 350.

Diez, P., Diez, N., Morrondo, P., & Cordero, M. (1990). Broncho-pulmonary helminths of chamois (*Rupicarupicapra parva*) captured in north.west Spain: assessment from first stage larvae in faeces and lungs. *Ann. Parasitol. Hum. Com.*, **2**, 74 – 79.

Donat, F., Ducos-de-Lahitte, J., Ducos-de-Lahitte, B., & Alcouffe, T. (1989). Contribution a l'etude des helminthes de l'isard (*Rupicapra pyrenaica pyrenaica*) survi epidemiologique de deux population dans le massif Roc Blanc-Carlit. *Gibier-Faune-Sauvage*, **6/6**, 383 – 402.

Dollinger, P. (1974). Contribution to the knowledge of the endoparasite faune of chamois (*Rupicapra rupicapra*) in Switzerland. *Zeitsch. Jagdwiss.*, **20**, 115 – 118.

Dunn, A.M. (1969). The wild ruminant as reservoir host of helminth infection. *Symp. Zool. Soc. London.*, **24**, 221-248.

Erhardová, B., & Ryšavý, B. (1953). Effect of environment on pre-invasive stages of lung worm *Muellerius capillaris* Müller, 1889. *Čsl. Biol.* **2**, 33-36. (In Czech)

Erhardová, B., & Michálek, J. (1956). To the problems of transmission of parasitic worms between different ruminant species. *Veterinářství*, **2**, 3. (In Czech)

Erhardová, B., & Ryšavý, B. (1954). Reservoirs of parasitic infections of domestic animals in nature. *Zborník z parazitologické konference*, Bratislava SAV, 220-229. (In Czech)

Erhardová, B. (1957). Lung worms of our ruminants. *Věst. Čsl. Zool. Společ.*, **21**, 148-158. (In Czech)

Erhardová, B. (1962). Contribution to ecology of parasitic worms in ruminants. *Čsl. Parazitol.*, **9**, 191-199. (In Czech)

Erhardová-Kotrlá, B., & Ryšavý, B. (1967). Species diversity and specificity of lung and intestinal worms in domestic animals and game in CSSR. *Vet. Med.*, **12**, 697-702. (In Slovak)

Euzéby, J. Diagnostic experimental des helminthoses animales. Paris, 1958, 88.

Fuschelberger, H. Das Gamsbuch. Naturgeschichte, *Hege und Jagd des Gams und atwas von fainer Umwelt*, 2, 1939. F.C. Maper Verlag, München.

Futák, J. (1955). How to restore the Tatras nature. *Tatranské noviny*, **2**, 16, 2. (In Slovak)

Gašinec, I. Eliminating the impact of negative anthropogenic factors on decrease in the Tatra Chamois population in Tatras National Park. In Chamois protection. (eds Janiga M., Švajda J.), 2002, p. 217-228. Publ. TANAP, NAPANT, IHAB, Tatranská Štrba, Banská Bystrica, Tatranská Javorina.

Gebauer, O. (1932). Zur Kenntnis der Parasitenfauna der Gemse. *Zeitsch. Parasit.* **4**, 147-219.

Genchi, C., Manfredi, M. T. Roncaglia, R., Sioli, C., & Traldi G. (1982). Gastrointestinal helminths of wild ruminants: observations on the chamois (Rupicapra rupicapra L.) in the Val Belviso reserve. *Parassitologia,* **24**, 197-203.

Genchi, C., Manfredi, M.T., & Sioli, C. Les infestation naturelles des chévres par les strongyles pulmonaires en milieu Alpin. Les maladies de la chevre, Niort (France), **9 - 11** octobre, 1984, INRA 28, 347 – 352.

Genchi, C., Bossi, A. & Manfredi, M.T. (1985). Gastrointestinal nematode infections in wild ruminants Rupicapra rupicapra and Dama dama: influence of density and cohabitation with domesticruminants. *Parassitologia,* **27**, 211-23.

Glawischnig, W., Khaschabi, D., Schöpf, K., & Schönbauer, M. (2000). Ein seuchenhafter Ausbruchvon Salmonella enterica Serovar Dublinbei Gemsen (*Rupicapra rupicapra*) *Wien. Tierärztl. Monat,* **87**, 21–25.

Grubb, P. Family Bovidae. In: Wilson D.E., Reeder D.M. (eds) Mammal speciesof the world. A taxonomic and geographic reference. Washington (DC): Smithsonian Institution Press., 1963, pp. 393-414.

Guoth, S. (1960). Helminthofauna of Alpine Ibex (*Capra ibex*) and wild goat (*Capra aegagrus*) from the Tatras National park . *Biológia,* **25**, 421-426. (In Slovak)

Gurčiani, K.R. (1967). K izučeniju legočnych namatod dikich žvačnych životnych Gruzii. Soobšč. AN Gruz. *SSR,* **24**, 1, 227-232.

Hadač, E. (1960). Notes on the chamois and deer diet in the Seven Spring valley in the Belianske Tatras. *Zborník prác o TANAP,* **4**, 257-262. (In Slovak)

Hanzák, J., & Veselovský, Z. Through the Animal World. Savci I., Albatros, Praha,1960, 471-475. (In Czech)

Hell, P., & Chovancová, B. (1995). Current status and prospects of Tatra chamois *Rupicapra rupicapra* in Slovakia. *Folia venatoría,* **25**, 167 – 174. (In Slovak)

Houdek, I., & Bohuš, I. (1976). The story of the Tatras. *Šport,* Bratislava, 241. (In Slovak)

Hörning, B. (1975). Die rolle des Parasitenbefalls in den Wildbeständen. *Schweiz. Z. Forstwesen.,* **5**, 361-372.

Hörning, B., & Wandeler, A. (1968). Der Lungenwurmbefall von Reh und Gemse in einigen Gebieten der Schweiz. *Rev. Suis. Zool.,* **75**, 597-608.

Hugonnet, L., & Euzéby, J. (1980). La parasitisme chez les jeunes chamois de la réserve naturelle des Bauges. *Bull.' Acad. Vet. France ,* **53**, 77 – 85.

Hugonnet, L., Montagut, G., & Euzéby, J. (1981). Incidences réciproques des infestations helminthiques des ruminants sauvages et des ovins domestiques en alpage en Vanoise. *Bull. Soc. Sci. Vet. Med. Com. Lyon,* **83**, 193 – 199.

Chovancová, B. Effect of human activities on the threatened vertebrate species in the TANAP, Tatras chamois, Tatras marmot and Golden Eagle. *Final report,* TANAP, 1985 Tatranská Lomnica.

Chovancová, B. Current status and prospects for preserving selected threatened vertebrate species in the TANAP. *Final Report. TANAP Research Centre,* 1990a, Tatranská Lomnica.

Chovancová, B. The possibility of the reducing negative anthrophic factors on selected threatened vertebrate species in the TANAP. *Final Report. TANAP Research Centre,* 1990b, Tatranská Lomnica.

Chovancová, B., & Šoltésová, A. (1988). Trophic base of Tatra chamois (*Rupicapra rupicapra tatrica* Blahout 1971) in the TANAP. *Folia venatoria,* **18**, 307-315. (In Slovak)

Chudík, I. (1968). Chamois in the Tatra National Park. *Myslivost,* **3**, 55. (In Slovak)

Chroust, K. (1991). Parasite fauna of chamois in Jeseníky. *Folia venatoria,* **21**, 77 – 88. (In Slovak)

Chudík, I. (1969). On the causes of decrease in chamois population in the TANAP. *Ochrana fauny,* **3**, 75-88. (In Slovak)

Chudík, I. (1974).Causes of losses and impact of large carnivores on hoofed game in the TANAP. *Folia venatoria,* **4**, 83-94. (In Slovak)

Janiga M., Hrkľová G. Chamos (*R. rupicapra*) in the diet of wolves (*C. lupus*) in the Tatra Mountains and the Low Tatra Mountains. In Chamois protection. (eds Janiga M., Švajda J.), 2002, p. 51-55. Publ. TANAP,NAPANT, IHAB, Tatranská Štrba, Banská Bystrica, Tatranská Javorina.

Janiga, M., & Zámečníková, H. Zoological characteristics of the historical data on chamois (*Rupicapra rupicapra tatrica,* Blahout, 1971) as a base for the evaluation of their current abundance in Tatra mountains. In Chamois protection. (eds Janiga M., Švajda J.), p.101-182, 2002. Publ. TANAP, NAPANT, IHAB, Tatranská Štrba, Banská Bystrica, Tatranská Javorina.

Klimaszewski, M. Geomorphology. In Przyroda Tatrzanskiego parku narodowego (eds. Z. Mirek, Z. Glowacinski, K. Klimek, H. Pienkos-Mirkova), 1996, p. 97-124. TPN, Krakow, Zakopane. (In Polish)

Kmeť, V., Pristaš, P., Janošková, A., & Janiga M. Comparison of the nucleotide sequences of gene for cytochrom b in chamois. In Chamois protection. (eds Janiga M., Švajda J.), 2002, p.35-39. Publ. TANAP, NAPANT, IHAB, Tatranská Štrba, Banská Bystrica, Tatranská Javorina.

Knaus, W., & Schröder, W. Das Gamswild. Verlag Paul Parey, 1960, Hamburg-Berlin, 206-208.

Kotrlý, A. (1958). Lung helminthic fauna of hoofed game in CSR. *Čsl. Parazitol.*, **2**, 101-110. (In Czech)

Kotrlý, A. (1962). Parasites of chamois in the Jeseníky region. Ústav vědecko-technických informací MZLVH. *Lesnictví,* **8**, 941. (In Czech)

Kotrlý, A. (1964). Ecology of parasites of hoofed game (Cervidae and Bovidae) in CSSR. *Práce výskumních ústavu lesníckych* ČSSR, **29**, 7-47. (In Czech)

Kotrlý, A. (1967). Relationship between parasites and hoofed game and domestic animals. *Vet. Med.*, **12**, 745-752. (In Czech)

Kotrlý, A., & Kotrlá, B. Helmithofauna of chamois (*Rupicapra rupicapra*) from Jeseniky and the Lužicke mountains in Czechoslovakia (ČSR) Práce VÚLHM, **39**, 1970, 59-77. (In Czech)

Kotrlý, A., & Kotrlá, B. (1977). Intestinal parasitic worms of hoofed game in Czechoslovakia. *Folia parasitologica,* **24**, 35-40. (In Czech)

Kotrlá,B.,Černý, V., Kotrlý, A., Minář, J., Ryšavý, B., & Šebek Z. *Parasitoses of game*. Academia Praha, 1984, 191 p.

Košťalik, J. Geographical character of the Spiš region since Eemian interglacial to this day (in relation to paleolithic settlement). In Spiš v kontinuite času (ed. P. Švorc), 1995, 299-308. Universum Prešov.

Kováč, J. The history of the care and protection of the Tatra Chamois (*Rupicapra rupicapra tatrica* Blahout, 1971) in the TANAP. In Chamois protection. (eds Janiga M., Švajda J.), 2002 p.197-204. Publ. TANAP,NAPANT, IHAB, Tatranská Štrba, Banská Bystrica, Tatranská Javorina

Krause, M., & Schmidt K. Food or safety – what comes first differs between male and female chamois *Rupicapra rupicapra* L. *Volume of Abstracts of the 2nd World Conference on Mountain Ungulates,* 5 – 7th May 1997, s. 88-89. Saint Vincent, Aosta.

Kratochvíl J., & Bartoš, E. Terminology and Classification system of Animals. Naklad. ČSAV Praha, 1954. (In Czech)

Krokavec M. Jr., & Krokavec M. Sr. (1991). Helminth fauna of Alpine chamois in the Slovak Paradise. *Veterinářství*, **41**, 3 – 4. (In Slovak)

Krupicer, I., Juriš, P., Dubinský, P., Vasilková, Z., Papajová, I., Štefančíková A., & Chovancová B. Prevalence of intestinal nematodes in the Tatra Chamois (*Rupicapra rupicapra tatrica*) in the TANAP-e in 1997. In Hygienic and ecological problems in relation to Veterinary Medicine,1998, 65-71. Publ. ŠVS, UVL, ÚEVM, PAU SAV. (In Slovak)

Ksiažek, J. Chamois in the Belianske Tatras - preliminary results on chamois population from monitoring the area and proposals of methods and organization of 'one time' overall inventory bz an example of the Belianske Tatras. InChamois protection. (eds Janiga M., Švajda J.), 2002, p. 229-236. Publ.TANAP, NAPANT, IHAB, Tatranská Štrba, Banská Bystrica, Tatranská Javorina.

Kutzer, E., & Hinaidy, H.K. (1969). Die Parasiten der wildlebenden Wiederkäuer Österreichs. *Z. Parasitenk.*, **32**, 354 - 368

Lovari, S. Behaviour as an Aid to Systemtematics. The case of the Abruzzian Chamois *Rupicapra rupicapra ornata* Neumann 1899. In Abstracts17th Ethiological Conference, 1981, Oxford.

Lovari, S., & Locati, M. (1991). Temporal relationships and structure of the behavioural repertoire in male Appenine Chamois during the rut. *Behaviour*, **119**, 77-103.

Lovari, S., & Scala, C. (1980). Revision of the Rupicapra genus. I. A statistical re-evaluation of Couturier`s data on the morphometry of six chamois subspecies. *Bull. Zool.*, **47**, 113-124.

Ložek, V. The Key to molluscs of Czechoslovakia. 1974. Vydav. SAV, Bratislava. (In Slovak)

Masini, F., & Lovari, S. (1988). Szstematics, phylogenetic-relationships and dispersal of the chamois (*Rupicapra* spp.) *Quat. Res.*, **30**, 339-349.

Midriak, R. Morphogenesis of the surface of high mountains, 1983, Veda, Bratislava. (In Slovak)

Michelčík, M. (1946). On the current state of chamois in the Tatras. *Poľovnícky obzor*, **14**, 132-135. (In Slovak)

Mituch, J. (1969). To the helminth fauna of hoofed ruminants in the TANAP. *Čsl. Ochrana prírody*, **8**, 237-250. (In Slovak)

Mituch, J. (1974). Helminth fauna of Aves and Mammalia. *Zborník prác a Tatranskom národnom parku*, **16**, 43 – 64. (In Slovak)

Mituch, J., Hovorka, J., Hovorka, I., & Tenkáčová, I. The Synthesis of obtained and general knowledge on helminths and helminthic cenosis of field game. *Final Report of the HELÚ SAV*, Košice, 1985, 101 p.

Mituch J., Hovorka, I., Hovorka, J., & Tenkáčová, I. Helminths and helminthic cenosis of large free-living mamals of the Carpatian Arch. *Final Report, HELÚ SAV*, Košice, 1989, 97 p.

Musil, R. (1985). Paleobiography of terrestrial counities in Europe during the Last Glacial. Sb. Národního Musea v Praze, B, *Přírodní vědy*, **41**, (1-2), 1-83 (In Czech).

Müller, A. (1889). Die Nematode der Säugethierlungen und die Lungenwurmkrankheit. *Tiermed. Verlg. Pathol.*, **15**, 261-321.

Mutafov, L., Bankov, D., Stoev, V., Youzev, P., & Karov, R. (1989). Experiments for therapy and chemical prophylaxis of gastric and intestinal and pulmonary strongylatoses in wild ruminantis.. *Vet. Sbirka* **5**, 40-42 (in Bulgarian).

Nešić, D., Pavlović, I., Valter, D., Mitić, G., & Hudina, V. (1972). Helminth fauna of chamois, muflon, reed deer, roe deer in Beograd. (Helmintofauna divikoza, muflona, jelena i srna u Beogradskom zoološkom vrtu tokom1990. godine.) *Vet. Glasnik* **46**, 92-101.

Nocture, M. Étude de l' infestation des pastureges l' altitude par les strongles des chamois. *These Doctor Veterinaire*, Lyon, 1986, pp. 109

Pavlovský, F. (1935). To the conservation of Tatra chamois. *Stráž myslivosti*, **1**, 7. (In Slovak)

Pavlovský, F. (1936a). Our chamois in the Tatras. *Stráž myslivosti*, **1**, 6-10. (In Slovak)

Pioz, M., Loison, A., Gibert, P., Dubray, D., Menaut, P., Le Tallec, B., Artois, M., & Gilot-Fromont, E. (2007). Transmission of a pestivirus infection in a population of Pyrenean chamois, *Vet. Microbiol.*, **119**, 19–30.

Pupkov, P.M. (1971). Helminth fauna of Rupicapra rupicapra caucasica in Central Caucasus. *Trudy Gorskogo Geľskohozyaistvennogo Inst. Yubileinyi.*, **32**, 331-333.

Radaelli, E., Andreoli, E., Mattiello, S., & Scanziani, E. (2007). Pulmonary actinomycosis in two chamois (*Rupicapra rupicapra*). *Eur. J. Wildl. Res.*, **53**, 231–234.

Radúch, J., & Karč, P. Current status and prospects of chamois population in the Low Tatras National Park. In: Current status and prospects of introduced populations of Tatra chamois in Slovakia. Publ. House of Technology ČSVTS Banská Bystrica, 1981 p. 7 – 30. (In Slovak)

Radúch, J. The importance of predators in the ecology of the Tatra chamois. In Chamois protection. (eds Janiga M., Švajda J.), 2002 p. 56-66.. Publ. TANAP, NAPANT, IHAB, Tatranská Štrba, Banská Bystrica, Tatranská Javorina.

Rajský, D., & Beladičová, V. (1987). Supplements to the studies on helminthoses and helminthic relations of chamois (*Rupicapra rupicapra* Linné, 1758) in the central part of the High Tatras. *Veterinářství*, **37**, 515 – 517.

Rodonaja, T.E. (1962). Materialy k izučeniju geľmintofauny dikich žvačnych Gruzii. *Soobšč. AN Gruz. SSR*, **27**, 709-716.

Rodrigues, F., Hamme,r S., Pérez, T., Suchentrunk, F., Lorenzini, R., Michallet, J., Martinková, N., Alboronz, J., Dominguez, A. (2009). Cytochrome b phylogeogrphy of chamois (*Rupicapra* spp.). Population contractions, expansions and hybridizations governed the diversification of the genus. *J. Heredity*, **100**, 47-55.

Ruchljadev, D.P. (1950). Legočnaja nematoda *Neostrongylus linearis* (Marotel, 1913) u dikich žvačnych Kaukaza. *Trudy GELAN*, **4**, 133-135.

Salzman, H.C., & Hörning, B. (1974). Der parasitologische Zustand von Gemspopulationen des Schweizarischen Juras im Vergleich zu Alpengemsen. *Z. Jagdwiss.*, **20**, 105-115.

Sattlerová-Šefančíková, A. Study on the ontogenetic stages of lung chamois biohelmints in the TANAP. PhD. Thesis. Helminthological Institute of SAS, Košice, 1981, 164 pp.

Sattlerová-Štefančíková, A. (1982). The resistance of first stage larvae of *Muellerius* spp. and *Neostrongylus linearis* (from the feces of chamois, *Rupicapra r. tatrica*) to different physical factors under laboratory and natural conditions. *Helminthol.*, **19**, 151 – 160.

Sattlerová-Štefančíková, A. (1987). Ecological conditions for lungworm infections of chamois in the Tatra National Park. *Biológia* (Bratislava), **42**, 113 – 119.

Sattlerová-Štefančíková, A. *Chamois and its parasites*. PressPrint, Košice, 2005, 124 p.

Sekera, J. (1938). Chamois of the High Tatras. *Naší Přírodou*, **23**, 882-885. (In Slovak)

Schmid, N. Deplazes, P., Hoby, S., Degiorgisb, M. P. R., Edelhoferc, R., & Mathis, A. (2008). *Babesia divergens*-like organisms from free-ranging chamois (*Rupicapra r. rupicapra*) and roe deer (Capreolus c. capreolus) are distinct from B. divergens of cattle origin – An epidemiological and molecular genetic investigation.*Vet. Parasitol.* **154**, 14-20.

Siko, S.B., & Negus, S. (1988). Aspects of the interrelations between parasitofauna of chamois (*Rupicapra rupicapra carpatica*, Couturier, 1938) and sheep (*Ovis aries* L.) from the same trophic areas. Erkrankungen der zootiere. *Verhandlungsbericht des 30 International Symposium über die Erkrankungen der Zoo und Wildttiere Vonn.*, **11**, 139-148.

Sokol, J. (1970). Chamois (*Rupicapra rupicapra*) mortality in the Kriváň region and problems of preserving chamois populaiton in the TANAP. *Ochrana fauny*, **4**, 49-64. (In Slovak)

Strnádová, J. The Diet of the Grey wolf (*Canis lupus* L., 1785) in the Slovak Carpathians. In *Chamois protection*. (eds Janiga M., Švajda J.), 2002, p. 45-49. Publ. TANAP,NAPANT, IHAB, Tatranská Štrba, Banská Bystrica, Tatranská Javorina. (In Slovak)

Stroh, G. (1936). Lungenwurmfunde bei 100 Gemsen und ihre Krankmachende Bedeutung. *Berl. Tierärz. Wschr.*, **43**, 696-699.

Šofránková, M. Gastropods of the Slovak Paradise National Park. Undergraduate Thesis, UPJŠ, Košice, 1982, 1-39. (In Slovak)

Špeník, M. Anthropozoonoses in free-living animals in a purpose-built centre of VŠV, the TANAP and PZ. *Final Project Report No. VI -6 - 2/9, VŠV Košice*, 1975, 122 p . (In Slovak)

Štefančíková, A. (1994). Lung nematodes of chamois in the Low Tatra National Park, Slovakia. J. *Helminthol.* **68**, 347 – 351.

Štefančíková, A. (1999). Lung nematodes in chamois (*Rupicapra rupicapra rupicapra*) of the Slovak Paradise National Park. *Acta Parasitol.*, **44**, 4, 255-260.

Štefančíková, A., Chovancová, B., Dubinský P., Tomašovičová, O., Čorba, J., Königová, B., Hovorka, I., & Vasilková, Z. (1999a). Lung nematodes of chamois (*Rupicapra rupicapra tatrica*, from the Tatra National Park, Slovakia. *J. Helminthol.*, **73**, 259-263.

Štefančíková, A., Chovancová, B., Dubinský P., Tomašovičová, O., Čorba, J., Königová, B., Hovorka, I., & Vasilková Z. (1999b). Current status of occurrence of lung nematodes in Tatra chamois (*Rupicapra rupicapra tatrica*, Blahout, 1971) in the Tatras National Park (TANAP). *Štúdia TANAP-u* , **4**, 179-188. (In Slovak)

Štefančíková, A., Vasilková, Z., Krupicer, I., Chovancová, B., Dubinský, P., Tomašovičová, O., Čorba, J., & Königová A.: Parasitoses of chamois and deer in the Tatras National Park. In Chamois protection. (eds Janiga M., Švajda J.), 2002, p.15-22. Publ. TANAP, NAPANT, IHAB, Tatranská Štrba, Banská Bystrica, Tatranská Javorina.

Šteffek, J. Molluscs of Slovak Paradise *Diploma Thesis*, PF UK Bratislava, 1975, 35 pp.

Švarc, R. Pathomorphological picture in lung nematodoses in chamois. *Final Report, HELÚ SAV*, 1982, Košice, p. 71.(In Slovak)

Švarc, R. (1984). Pulmonary nematodes of the chamois *Rupicapra rupicapra tatrica*, Blahout, 1971. I. Pathomorphological picture of lungs during the development of worms into the adult stage. *Helminthol.*, **21**, 141-149.

Teren, Š. (1957). Chamois and its conservation. *The Hunting Horizon,* 4, 82-88. (In Slovak)

Thomson, E.F., Gruner, l., Bahhady, F., Orita, G., Termanini, A., Ferdawi, A.K., & Hreitani, H. (2000). Effects of gastro-intestinal and lungworm nematode infections on ewe productivity in farm flocks under variable rainfall conditions in Syria. *Livest. Prod. Sci.,* 63, 65-67.

Venátor, J. Chamois hunting in the Liptovské hole. *Naša zver.* Academia Publishing House, Bratislava, 1935, 113-122. (In Slovak)

Vichová, B., Majláthová, V., Nováková, M., Bullová, E., Štefančíková A., & Peťko, B. The role of wild animals in the maintenance and circulation of *Anaplasma phagocytophilum* in the natural conditions of central Europe. In *Xth European Multicolloquium of Parasitology, Program and Abstracts Book,* Paris, France, August 24th -28th, 2008, , p. 197.

Voskár, J. (1993). Ecology of the Grey wolf (*Canis lupus*) and its share on formation and balance of the Carpathian ecosystems in Slovakia. *Ochrana prírody,* 12, 243-276. (In Slovak)

Zelina, V. (1956). Tatra chamois in winter. *Myslivost,* 1, 22-23. (In Slovak)

Zemanová, B, Hájková, P, Bryja, J, Mikulíček, P, Martínková, N, Hájek, B., & Zima J (2007). Conservation genetics of chamois populations in Slovakia. In: Prigioni C & Sforzi A (Eds). Abstracts V European Congress of Mammalogy. *Hystrix It J Mamm* (ns) Vol 1-2, Suppl (2007), p 28.

In: National Parks: Biodiversity, Conservation and Tourism
Editors: A. O'Reilly and D. Murphy, pp. 117-130
ISBN: 978-1-60741-465-0
© 2010 Nova Science Publishers, Inc.

Chapter 6

THE DILEMMA OF BALANCING CONSERVATION AND STRONG TOURISM INTERESTS IN A SMALL NATIONAL PARK: THE CASE OF AMBOSELI, KENYA

Moses Makonjio Okello, John W. Kiringe and John M. Kioko
The School for Field Studies, Center for Wildlife Management Studies, Nairobi, Kenya

Abstract

Amboseli is a small park located in the wildlife rich area of Tsavo-Amboseli Ecosystem in Southern Kenya. It is one of the leading parks in terms of absolute tourism revenue generated per year and one of the highest in terms of tourism revenue per unit conservation area. Its revenue is sometimes in excess of Ksh. 100 Million (US$1.33 Million) per year. However, as important as this park is to national and local economy, it has many challenges to its viability in terms of balancing conservation and tourism interests. Several reasons explain this. Its gazettement as a national park in 1974 did not carefully consider implications for the small size in light of its critical role as a dry season wildlife refuge. It is only a fragment of a formerly larger ecosystem, which has now been taken over by human settlement and other incompatible land uses to wildlife conservation. With mainly tourism revenue as a motivation, the government displaced the Maasai from the park and promised to provide them with water from Amboseli swamps or they could water their livestock in the park if this agreement failed. This historical agreement together with rising elephant population, increasing management and tourism infrastructure in the park, the contraction and land use changes in Amboseli Maasai dispersal areas and combine to make Amboseli National Park an increasingly unsustainable conservation unit. This paper discusses the dilemma of managing Amboseli for biodiversity conservation and as a popular tourist destination.

Key words: Amboseli, biodiversity, conservation, Kenya, sustainable tourism

Introduction

Amboseli is one of the fifty-three protected areas in Kenya covering about 8% of the countries landscape (Nyeki 1992). Located in southern western Kenya, it was gazetted in

1974 as a national park after existing from early 1960's as a big (over 8,000 km^2) game reserve. It covers an area of 392 km^2. Located close to Mt. Kilimanjaro, it is one of the three national parks that form the expansive Tsavo – Amboseli Ecosystem which includes Chyulu Hills National Park, Tsavo West National Park, and six group ranches (Mbirikani, Kuku, Kimana, Rombo, Olulugui – Ololorashi, and Eselengei). The group ranches are communally owned by the Maasai people and forms a critical dispersal area for neighboring national parks' wildlife, as well as resident area for free ranging wildlife (Wishitemi & Okello 2003, Okello *et al.* 2003).

The historical agreement that led to the transformation of Amboseli from a national reserve to a national park between the Maasai and the government seem to have heralded the current challenges to conservation of wildlife in the Amboseli Ecosystem. It has been one of the causes for park degradation as livestock also concentrates in the park in the dry season for pasture and water. The small park size has been unable to support the ever-increasing elephant numbers, with about 11,000 elephants instead of about 200 that the park can support. The result has been loss of woodlands, thereby reducing the parks' species diversity.

Elephant local over - population is also a major cause for human-wildlife conflicts. Elephants have become a leading problem animal in crop raiding, killing and injuring both livestock and people. The three large tourist lodges located close to critical swamps that support large ungulates in the dry season. When combined with management facilities (ranger accommodations, administration offices and gates, and an airstrip) a high road network and tourist density, this puts more pressure on Amboseli by further reducing further the effective conservation area available to wildlife.

All these factors combine to make Amboseli National Park an increasingly unsustainable conservation unit and unlikely to contribute tourism revenue to national treasury in the future. We argue that the viability of Amboseli will require these issues to be addressed and a clear policy dealing with balancing conservation and economic issues be formulated. This chapter looks at these challenges and the way forward. Insights from this paper will be useful for managing highly visited but small insularized national parks in East Africa.

Tourism and Revenue Contribution of Amboseli

Amboseli National Park is third to Nakuru and Nairobi national parks in terms of absolute number of tourists received per year (Table 1) and possibly revenue generated among protected areas in Kenya. It is currently the most visited park in Kenya. These three parks are relatively similar in their contribution to tourism revenue generation to the government. They are therefore valuable in providing the foreign revenue for government. Amboseli, in particular, is the sixth highest tourism earners in terms of revenue per unit area of conservation. This contribution makes it a valuable asset to the government and to the tourism industry. It based on this that the Kenyan government in 1974 gazetted this area covered by permanent swamps that attracted a high density of large ungulates especially during the try season. This concentration of large ungulates is what has brought in tourists to film, photograph and observe over the years.

Table 1. Leading protected areas in Kenya based on mean tourist numbers per unit conservation area among protected areas in Kenya (adopted from Okello & Kiringe 2004).

Protected area	Designation Date	Agency Responsible	Area (km^2)	Mean annual tourist numbers (over 12 years from 1989) Mean ± SE (rank)	Mean tourist numbers per unit area of conservation (rank)
Nairobi Park	1946	KWS	117	147,072.7 ± 5 125.52 (2)	1,257.02 (1)
Kisite – Mpunguti Marine	1978	KWS	28	29,381.82 ± 1956.50 (12)	1089.34 (2)
Lake Nakuru Park	1961	KWS	188	154,918.2 ± 5 196 (1)	824.03 (3)
Hellsgate Park	1984	KWS	68	39,918.18 ± 3922.17 (9)	587.03 (4)
Watamu Marine	1968	KWS	42	22,009.09 ± 1636.84 (15)	524.02 (5)
Buffalo Springs Reseve	1985	KWS	131	68,327.27 ± 641.55 (7)	521.58 (6)
Amboseli Park	1974	KWS	392	141,581.8 ±1 331.09 (3)	361.18 (7)
Lake Bogoria Reserve	1970	Local Authority	107	34,463.64 ± 2476.11 (11)	322.09 (8)
Malindi Marine	1968	KWS	219	35,636.36 ± 2476.11 (10)	162.72 (9)
Shimba hills Reserve	1968	KWS	192	27,945.45 ± 3610.37 (13)	145.55 (10)
Mombasa Marine	1986	KWS	210	27,272.73 ± 3610.37 (14)	129.87 (11)
Aberdares Park	1950	KWS	715	60,190.91± 1641.93 (8)	84.18 (12)
Maasai Mara Reserve	1961	Local Authority	1672	139,127.3 ± 7864 (4)	83.21 (13)
Kiunga Marine	1979	KWS	250	16,990.91± 319.72 (16)	67.96 (14)
Samburu Reserve	1985	KWS	165	6,131.82 ± 687.14 (19)	37.16 (15)
Mount Kenya Park	1949	KWS	715	15,872.73 ± 726.03 (17)	22.20 (16)
Meru Park	1966	KWS	870	9,518.18 ± 1762.03 (18)	10.94 (17)
Tsavo East Park	1948	KWS	11,747	127,463.6 ± 1169.92 (5)	10.85 (18)
Tsavo West Park	1948	KWS	9,065	85,890 ± 8777.7 (6)	9.47 (19)

The number of tourists per unit area of conservation is an index of congestion. National parks with more tourist density are generally more congested. This unit of measurement (tourists per unit area of conservation) distinquishes parks that receive many tourists over vast unused lands compared to those who receive many tourists yet their sizes are small. The later assessments reveal parks that may be congested and hence exceeded their tourist carrying capacity. Nairobi, Nakuru, Hellsgate and Buffalo springs top among the terrestrial parks while Kisite – Mpunguti Marine and Watamu top congested marine protected areas. However, since every marine area is a composite of park and reserve, the area given are for the national park rather than the reserve component of the protected area. If an inclusive area was available, they would be devalued in terms of tourists per unit area of conservation. Amboseli National Park is ranked as the fifth most congested national park in Kenya (table 1).

Table 2. Twenty leading protected areas in terms of tourist attraction scores (based on a maximum of five) and some of the major tourist attractions for Kenya's protected areas. (Source: Adopted from Okello et al. 2001)

Protected Area	"Big five" Large Mammals	Terrestrial Large Mammals	Physical features	Cultural attractions	Number of bird species	Species of conservation concern	Aesthetic rating	Mean score	Rank
Amboseli	5	5	3	5	5	4	3	4.29	1
Masai Mara	5	5	2	5	5	4	4	4.29	1
Tsavo East	5	5	4	3	5	4	4	4.29	1
Tsavo West	5	5	4	3	5	3	4	4.14	4
Buffalo Springs	4	5	3	4	4	4	4	4.00	5
Samburu	4	4	3	4	4	4	4	3.86	6
Hellsgate	3	2	5	4	4	4	5	3.71	7
Meru	4	5	3	2	4	4	4	3.71	7
Nairobi	4	3	3	3	4	4	3	3.43	9
Lake Nakuru	4	4	4	1	4	3	5	3.43	9
Kakamega	1	2	3	4	4	5	4	3.29	11
Kisite-Mpunguti	0	2	5	4	2	5	4	3.14	12
Mount Kenya	4	3	4	1	1	5	5	3.14	13
Longonot	3	2	2	4	3	2	5	3.00	14
Shaba	4	1	3	4	1	4	4	3.00	14
Marsabit	1	1	3	3	4	4	5	3.00	14
Mount Elgon	3	2	3	1	2	5	5	3.00	14
Sibiloi	3	2	3	3	1	4	4	2.86	18
Chyulu Hills	4	2	3	3	1	4	3	2.86	18

Since tourism in Kenya is mainly wildlife – based (Okello *et al.* 2001, Sindiga 1995), Amboseli National Park is leading in terms of tourist attractions. Okello *et al.* (2001) considered biological (*big five* mammals[1], large mammal species, bird species, and rare species), physical (physical features, landscape aesthetics), and cultural endowments in prioritizing protected areas in terms of tourism appeal. Amboseli National Park has a high density of animals that tourists can see easily. All the b*ig five* (other than the rhino) are present, including its proximity to Mount Kilimanjaro (the world largest and highest free standing mountain) and location in the world culturally renowned Maasai people. These authors ranked Amboseli at par with Maasai Mara and Tsavo West national parks as leading in terms of tourist attractions. This explains why it is one of the highest tourism earners in both absolute revenue per protected area, revenue per unit area of conservation and tourist congestion in the country.

But tourist attractions are not the only parameters that determine tourism potential (Okello *et al.* 2001). There are parks with equally spectacular or more diverse tourist attractions but do not get many tourists. Their potential is hindered by other factors such as marketing, security, communication infrastructure, and quality of facilities and ease of access. Okello *et al.* (In Press) has assessed the potential of protected areas in Kenya not only in terms of tourist attractions (biological, physical and cultural) but also in terms of marketing campaigns, risks (such as crime and health) and facilities (such as communication and accommodation). When all these factors influencing tourism are considered (table 3),

---

[1] Lion (Panthera leo), Leopard (Panthera pardus), Elephant (Loxodonta africana), Rhinoceros (Diceros bicornis), Buffalo (Sincerus caffer).

Amboseli National Park ranks the top protected area in Kenya. Its ability to attract tourists includes world-class available accommodation, ease of access and proximity to the cities of Nairobi and Mombasa, which are international entry points of tourists into Kenya. This underscores its importance as a conservation area as well as a tourist destination. It is endowed with tourist attractions, communication and accommodation facilities, and well marketed internationally through documentaries.

Table 3. Top twenty protected areas based on the mean scores of the six factors influencing tourism potential in Kenya (Adopted from Okello et al. 2005).

Protected area	Biodiversity attributes	Physical Attributes	Cultural Attractions	Marketing Campaigns	Number of facilities	Health and insecurity	Mean score	Ranks based on mean scores
Amboseli	4.29	3.00	5.00	4.50	3.29	5.00	4.18	1
Hellsgate	2.51	5.00	4.00	4.50	3.71	4.00	3.96	2
Maasai Mara	4.00	3.00	5.00	4.50	3.43	3.00	3.82	3
Longonot	3.14	3.50	4.00	3.71	3.00	4.50	3.81	4
Tsavo East	4.00	4.00	3.00	4.00	4.00	3.00	3.67	5
Samburu	3.71	3.50	4.00	3.50	3.14	4.00	3.64	6
Nairobi	3.71	3.00	3.00	3.50	3.86	3.50	3.60	7
Tsavo West	3.86	4.00	3.00	4.00	3.57	3.00	3.57	8
Watamu Marine	2.00	4.00	4.00	3.50	4.43	3.50	3.57	8
Mount Kenya	3.43	4.50	1.00	5.00	3.57	4.50	3.50	10
Buffalo Springs	3.86	3.50	4.00	2.50	3.14	4.00	3.50	10
Malindi Marine	1.86	3.00	4.00	4.00	4.43	3.50	3.46	12
Lake Nakuru	3.71	4.50	1.00	4.50	3.57	4.00	3.38	13
Kisite – Mpunguti	2.43	4.50	4.00	3.00	3.86	2.50	3.38	13
Mombasa Marine	1.86	2.50	4.00	4.00	4.43	3.50	3.38	13
Maralal Sanct	2.14	2.50	5.00	3.00	3.00	4.00	3.27	16
Shaba	2.86	3.50	4.00	2.50	2.71	4.00	3.26	17
Mount Elgon	2.71	4.00	1.00	5.00	3.14	3.50	3.22	18
Aberdares	3.00	3.00	1.00	4.50	3.86	4.50	3.14	19
Kakamega	3.00	3.50	4.00	2.00	2.86	3.50	3.14	19

Okello et al (2005) uses an index to measure the level of tourism achievement of protected areas in Kenya based on actual tourism numbers and the potential in terms of attractions and enabling environment for tourists (table 4). This achievement index (referred to as tourism achievement index, *tai*) shows that Amboseli National Park (among other small congested protected areas in Kenya) has achieved and or exceeded its tourism potential and cannot continue to accommodate tourists beyond the average numbers it has been receiving. Doing so will be at the detriment of its conservation objectives and a danger to the biodiversity it safeguards (Johnstone, 2000). It is important to keep a balance between the number of tourists and the physical environment that can support it. The danger of negative tourism impacts (such as animal harassment, off – road driving, feeding wildlife, congestion of tourist lodges and vehicles, and over – speeding) threaten long term viability of such small parks that have exceeded their tourism carrying capacity. Therefore a viable strategy is to maintain tourism by targeting few but high – paying tourists. What is now required is clearly an estimate of visitor carrying capacity and examine impacts to physical environment and animals behavior and park use when this level is exceeded. But as this and other evidence below reveals, tourism is not in balance with conservation objectives in Amboseli National Park.

Table 4. Tourism achievement based on tourist numbers and potential (calculated from biodiversity value) for leading protected areas in Kenya

Protected area	Date of designation	Tourist number, rank and scores (S)	Tourism Potential indices (tai index for achievement)	Remarks on tourism potential
Aberdares	1950	60190.91 (S= 5)	3.14 (19) (tai = 0.65)	Exceeded tourism potential
Lake Nakuru	1961	154918.2 (S= 5)	3.38 (13) (tai* = 0.57)	Exceeded tourism potential
Buffalo Springs	1985	68327.27 (S= 5)	3,50 (10) (tai = 0.53)	Exceeded tourism potential
Tsavo West	1948	85890 (S= 5)	3.57 (8) (tai = 0.50)	Exceeded tourism potential But can take more due to bigger size
Nairobi	1946	147072.7 (S= 5)	3.60 (7) (tai = 0.49)	Exceeded tourism potential
Tsavo East	1948	127463.6 (S= 5)	3.67 (5) (tai = 0.47)	Exceeded tourism potential But can take more due to bigger size
Maasai Mara	1961	139127.3 (S= 5)	3.82 (3) (tai = 0.41)	Exceeded tourism potential
Amboseli	1974	141581.8 (S = 5)	4.18 (1) (tai = 0.29)	Achieved tourism potential
Hellsgate	1984	39918.18 (S= 4)	3.96 (2) (tai = 0.01)	Achieved tourism potential But could take more tourists
Malindi Marine	1968	35636.36 (S= 4)	3.46 (12) (tai = 0.19)	Achieved tourism potential But could take more tourists
Lake Bogoria	1970	34463.64 (S= 4)	3.05 (22) (tai = 0.33)	Achieved tourism potential But could take more tourists
Kisite – Mpunguti Marine	1978	29381.82 (S= 3)	3.38 (13) (tai = -0.13)	Not achieved tourism potential
Shimba hills	1968	27945.45 (S= 3)	2.69 (28) (tai = 0.11)	Achieved tourism potential But could take more tourists
Mombasa Marine	1986	27272.73 (S= 3)	3.38 (13) (tai = -0.13)	Not achieved tourism potential
Watamu Marine	1968	22009.09 (S= 3)	3.57 (8) (tai = -0.20)	Not achieved tourism potential
Impala Sanctuary	1985	19222.73 (S= 2)	2.64 (30) (tai = -0.22)	Not achieved tourism potential
Kiunga Marine	1979	16990.91 (S= 2)	2.14 (47) (tai = -0.05)	Not achieved tourism potential
Mount Kenya	1949	15872.73 (S= 2)	3.50 (10) (tai =- -0.53)	Not achieved tourism potential
Meru	1966	9518.18 (S = 1)	3.13 (21) (tai = -0.75)	Not achieved tourism potential
Samburu	1985	6131.82 (S= 1)	3.64 (6) (tai =- 0.93)	Not achieved tourism potential

* Tai index: over 0.4 -- Exceeded tourism potential; 0 – 0.40 -- Achieved potential; Negative tai --Not realized potential.

Nature of Threats to Amboseli Biodiversity

Most protected areas in Kenya are threatened by a diversity of factors (Mwale 2000, Okello & Kiringe 2004). Amboseli National park is the forth most threatened protected area in terms of prevalent of threats to its biodiversity and that of its wildlife dispersal area (table 5). It has a prevalent index of 0.7 and is threatened by incidences of human encroachment, land – use changes, over – exploitation of natural resources, bush meat poaching, significant negative tourism impacts, loss of wildlife corridor and dispersal area, intense human – wildlife conflicts. Human population is increasingly encroaching around Amboseli National

Park and in the group ranches dispersal areas due to high natality as well as immigration from over – populated areas in Kenya and Northern Tanzania. (Campbell *et al.* 2000)

Table 5. Susceptibility of protected areas to threats against biodiversity and conservation within and around them as stated by protected area officers (Source: adopted from Okello and Kiringe 2004)

Protected Area	Threat factors within and outside a protected area	Protected Area Susceptibility Index (out of a max. of one
Masai-Mara National Reserve	1,2,3,4,5, 6,7,9, 10	0.9
Ndeere Island National Park	1,2,3, 4,5,6,9,10,	0.8
Lake Nakuru National Park	2,3,4,6,7,8,9,10	0.8
Amboseli National Park	2,3, 5,6,7,9,10	0.7
Aberdare National Park	1,2,5, 6,8,9,10	0.7
Mt. Elgon National Park	1, 2,3,5 ,6, 9,10	0.7
Kiunga Marine	1, 2,3,4, 5, 6,7	0.7
Mt. Kenya National Park	1,2, 3,5,6,9,10	0.7
Mombasa Marine	1,2,3, 4,5,6,7	0.7
Watamu Marine	1,2,3, 4,5,6,7	0.7
Ruma National Park	1,2, 3,5,6, 9,10	0.7
Kisitei-Mpunguti Marine	1, ,2,3,4,5, 6,7	0.7
Malindi Marine	1,2,3,4,5,6,7	0.7
Mwea National Park	1,2,4,6,8,9,10	0.7
Kamnarok National Reserve	1,2,3,5,6,9,10	0.7
Rimoi National Reserve	1,2,3,5,6,9,10	0.7
Nairobi N. Park	1,2,4,7,8,9,10	0.7
Tana River Primate National Reserve	1,2, 4,5,6,10	0.7
Ngai Ndeithya National Reserve	1,2,5,6,9,10	0.6
Saiwa Swamp National Park	1,2,3,5,6,9	0.6
Kakamega Forest National Park	1,2,5,6,9,10	0.6
Oldonyao-Sabuk National Park	2,3,4,6,9,10	0.6
Shimba Hills National Park	1,2,6,8,9,10	0.6
Tsavo-West National Park	1,2,5,6,9,10	0.6
Nasalot National Reserve	1,2,3,6,9,10	0.6
Chyulu National Park	1,2,5,6,9,10	0.6
South Turkana National Reserve	1,2,3,6,9,10	0.6

Key to Threats facing protected areas: 1- Poaching; 2 - Human encroachment; 3 - Land-use changes; 4- Pollution; 5 - Over-exploitation of natural resources; 6 - Bush meat; 7 – Negative tourism impact; 8 – Fencing; 9 - Loss of wildlife corridor and dispersal area; 10 - Human-wildlife conflicts.

Amboseli National Park may be more threatened than what the index show. It is a small park hived from a much larger ecosystem. As a result, it is unable to support high plant and wildlife biodiversity unless it has surrounding lands as dispersal areas. The surrounding group ranches are communally owned and not under jurisdiction of conservation agencies. Alternative land uses such as agriculture are expanding and degrading, converting and blocking the dispersal area and migration corridor between Amboseli / Kilimanjaro system and the Tsavo West / Chyulu system (Okello *et al.* 2003). With rising human wildlife conflicts due to opportunity costs, lack of wildlife compensation, wildlife – induced losses and lack of wildlife benefits to the local communities, support and tolerance for wildlife has greatly declined (Wishitemi & Okello 2003, Okello 2005). Community resentment is so high that retaliatory killing and snaring of wildlife is becoming common. As human encroachment increases, Amboseli Park is largely turning into an insularized system unable to meet the needs of long – ranging species. Human wildlife (especially elephant) conflicts and insularization are the biggest threats to Amboseli National Park. When this is combined by the commercially motivated tourism industry, future viability of Amboseli becomes a concern.

Whatever is remaining of the dispersal area of Amboseli within the group ranch is also under pressure for group ranch subdivision. Like any Maasai Group Ranches established in 1960's, sub – division has support of the Maasai and the government. It is already in progress in many group ranches (Campbell *et al.* 2000). The factor fuelling this subdivision is the desire for individual land that landowners can use to secure development loan, expand agriculture, invest in their own land and live a settled life. There are worries that land owned communally enlist little land stewardship, and can be grabbed by politically connected people as well as community elites. There are conservation related concerns that sub – division will negatively impact on wildlife by fragmenting the dispersal area and encouraging selling or fencing or any land use that is incompatible with pastoralism and wildlife conservation. Depending on whether individual owners will pool their land together so that they can still have blocks of land for pastoralism and allow access for wildlife, it is clear that the landscape under sub division will be very different from what they are today (Wishitemi & Okello 2003, Okello *et al.* 2003).

Elephant Impacts Consequences for Mammal Diversity

An examination of the park habitats clearly show the ever– declining Acacia woodland. Three broad habitats that have existed in Amboseli throughout its conservation history has been acacia woodlands, open grasslands and swamps. Pictures of Amboseli before 1980's show a typical savanna ecosystem with woodlands interspersed with grass glades supporting a diversity of mammals, grazers, browsers and mixed feeders among others (Western 2002). But today, open alkaline grasslands, comprising 63.73%, dominate the park. The woodland comprises only 5.33% while the swamps occupy 30.93% of Amboseli Park. This recent habitat changes (with increasing grassland and swamp areas) has led to Amboseli being dominated by plains game (plains zebra, common wildebeest, grants and Thomson gazelles and impala). The loss of woodlands is blamed on rising water table and expanding swamps, past off – road driving for this fragile environment, natural senescence of acacia woodlands, and finally the destructive actions of locally over – abundant elephants (Amboseli Warden Pers. Comm.).

We are convinced that the elephant damages greatly contribute to this decline. Most elephant researchers (Western 1989, White 2001) agree that an appropriate density for elephant should be one elephant per square mile (one to about 2.3km^2). Even for African parks that are concerned about local elephant overpopulation (such as Kruger National Park) has a density of less than one elephant per square kilometer. Recent surveys (table 6) show that Amboseli has over 1,160 elephants using its area over a year, especially in dry season.

Further, Amboseli elephant density is consistently over 2 elephants per square kilometer, with as high as four. This is one of the highest densities in Africa.

Table 6. Grazing guilds composition, elephants and other large mammal density (animals per km^2) and diversity in Amboseli National Park in dry season.

Large mammal attributes	Sub - attributes	Sept. 2001	Nov. 2001	Feb. 2002	June 2002	Sept. 2002	Feb. 2003	Sept. 2003
Elephant Density	-	4.33 ± 0.85	3.87 ± 2.28	2.49 ± 1.13	2.05 ± 0.73	2.16 ± 0.77	18.18 ± 8.48	2.57 ± 0.84
Wildebeest Density	-	121.37 ± 11.92	103.69 ± 29.48	3.84 ± 0.88	39.56 ± 7.18	53.72 ± 19.50	13.80 ± 2.52	40.42 ± 9.97
Zebra Density	-	84.89 ± 7.64	104.02 ± 62.36	20.93 ± 4.76	25.76 ± 5.27	29.29 ± 7.94	40.72 ± 14.50	34.61 ± 8.69
Thomson gazelle's density	-	47.87 ± 5.85	24.08 ± 6.73	4.28 ± 1.34	10.16 ± 2.29	7.71 ± 2.58	6.40 ± 1.38	9.28 ± 2.57
Grants gazelle		7.59 ± 0.82	5.66 ± 2.10	1.50 ± 0.50	5.94 ± 1.61	6.00 ± 2.70	3.39 ± 0.82	6.22 ± 1.75
Buffalo Density	-	5.48 ± 1.03	3.97 ± 1.96	2.70 ± 1.54	1.82 ± 1.39	6.28 ± 3.13	2.44 ± 1.13	2.75 ± 1.75
Impala	-	11.07 ± 2.20	4.02 ± 1.90	0.82 ± 0.79	1.54 ± 0.88	1.73 ± 0.87	0.75 ± 0.50	0.39 ± 0.20
Maasai giraffe	-	Not seen	Only 5 total in park	Not seen	0.23 ± 0.12	0.20 ± 0.11	0.36 ± 0.14	0.33 ± 0.15
Gerenuk	-	Not seen	Not seen	Not seen	Not seen	0.04 ± 0.02	Not seen	Not seen
Total large mammal species	-	18.14 ± 7.35	17.17 ± 6.34	2.49 ± 1.57	89.12 ± 13.42	108.76 ± 28.57	88.31 ± 18.64	98.57 ± 18.69
Livestock Density	-	3.24 ± 1.93	11.10 ± 8.71	1.46 ± 1.91	8.06 ± 4.04	21.38 ± 8.12	2.70 ± 1.21	15.47 ± 5.99
Large mammal species diversity and comparisons among the three main habitats of Amboseli National Park	Simpson Index, Ds	0.74(A)* 0.73(G) 0.70(S) 0.71(W)	0.77(A) 0.64(G) 0.63(S) 0.75(W)	0.81(A) 0.73(G) 0.82(S) 0.83(W)	0.76(A) 0.70(G) 0.72(S) 0.75(W)	0.74(A) 0.74(G) 0.66(S) 0.77(W)	0.78(A) 0.48(G) 0.75(S) 0.78(W)	0.73(A) 0.69(G) 0.70(S) 0.79(W)
	Statistical comparisons of Ds	G > S G = W S = W	G = S G < W S < W	G = S G = W S = W	G = S G < W S < W	G > S G = W S < W	G = S G < W S = W	G = S G < W S < W
	Inverse Simpson Index, ds	3.79(A) 3.77(G) 3.33(S) 3.42(W)	4.35(A) 2.77(G) 2.70(S) 4.00(W)	5.27(A) 3.71(G) 5.56(S) 5.89(W)	4.19(A) 3.35(G) 3.63(S) 4.03(W)	3.82(A) 3.90(G) 2.91(S) 3.90(W)	4.45(A) 1.92(G) 3.99(S) 4.45(W)	3.73(A) 2.62(G) 3.31(S) 4.70(W)
Herbivore feeding guilds proportions	Grazers	93.00 %	95.55 %	74.18 %	83.00 %	93 %	72%	79.25 %
	Browsers	1.00 %	4.44 %	0.00 %	2.00 %	1 %	0.40 %	0.36 %
	Mixed Feeders	6.00 %	0.005 %	25.82 %	14.00 %	6 %	25.30 %	20.01 %

* Key to the three main habitats of Amboseli National Park: **A** – Entire Amboseli Park; **G**—Grassland habitat; **S** – Swamp habitat; **W** – Woodland habitat
Data extracted from fieldwork by the first author, and manuscript in preperation.

This high density and compression (partly because of changing land uses outside the park and the presence of permanent water swamps in Amboseli) have led to obvious elephant impacts on woodland. The remaining woodland continues to die off due to direct elephant mortality or exposure to diseases and pathogens after elephants trip off the tree bark. Electric enclosures in the park that exclude elephants have instant acacia regeneration, supporting the view that high elephant density in Amboseli are responsible for the suppression of acacia woodland regeneration in Amboseli. As elephants destroy the woodlands, open grasslands expand, encouraging wildebeest, zebra, gazelles, impala and buffalo populations dominate Amboseli (table 6). These and other grazers comprise of over 70% of the park, with as high as 96% species composition. Browsers often comprise less than 1% of herbivore biomass in Amboseli.

The ecological role of elephants (e.g. in diversifying habitats and benefiting other species) can only be realized if appropriate elephant densities are maintained. This is clear when we examine the large herbivore diversity in the three habitats of Amboseli (table 6). The diversity of open woodlands (opened and maintained by elephants) is consistently higher than those of open alkaline grasslands or swamps. Therefore the destruction of woodlands has reduced habitat heterogeneity in Amboseli, resulting into permanent displacement or local extinction of browsers. Gerenuk (*Litocranius walleri*), lesser kudu (*Tragelaphus imberbis*), bushbuck (*Tragelaphus scriptus*), Eland (*Tragelaphus oryx*) and Masai giraffe (*Giraffa camelopardalis*) have been displaced (and mostly occur outside the park where woodlands still exist). When this is combined with habitat degradation and competition for forage by the Masai livestock at or near watering points, further stress on Amboseli Park becomes threatening to its future viability as a conservation area. Livestock, elephant and other plain herbivores densities cumulatively degrade Amboseli, sometimes with large mammal density of over 100 large mammals per square kilometer use the park. While this may be great for tourism, it is a constraint to its biodiversity conservation goals.

Impacts of Tourism and Management Infrastructure

The greatest rationale for conservation in Kenya, as a poor developing country, is tourism revenue. Since the ban on sport hunting and other wildlife utilization in 1977, tourism emerged as the best and most acceptable form of wildlife benefit. As such, the government strategy has always been to increase foreign tourist visits by investing (sometimes heavily in a few protected areas) in facilities that will enhance tourism. Amboseli National Park has roads, five star hotels and lodges, and entry gates all located within the park to serve the tourism industry. Management infrastructures such as accommodation for rangers and park headquarters have been constructed within the park. There is currently a road network of 250 km in Amboseli, taking an area of about 1.02 km^2 and (table 7). There are four gates with associated ranger accommodation facilities located at different areas of the park perimeter. A central area with two large lodges (with associated staff accommodation) is located next to a major wetland. Close to this lodge area is an airstrip, into which lands aircrafts carrying tourists, besides the busy gates. The other lodge is also located at the southern part of the park, also close to a second swamp (table 7). The park headquarters has machinery maintenance unit, ranger accommodation and administrative block located in the remnant acacia woodland habitat (figure 1).

All these enclosures, administrative and tourism structures comprise about 1.5% of the total park area. It may seem little sacrifice compared to the tourism revenue Amboseli generates, but when taken into the context of an insularized park in an increasingly resented environment, it becomes a viability concern. Further, its not only the physical size, but locations of these infrastructure. Lodges next to major swamps degrade and pollute swamps, hence strangling the very lifeline of Amboseli. These swamps are what Amboseli relies on to support high density of herbivores especially in the dry season. Now that they are there, they should be made to control effluents and garbage inside the park. There is so much human traffic together with high population of essential and non – essential workers in the lodge areas. The noise and other pollution they make negates the objectives of biodiversity conservation in the park. The same goes for ranger accommodation and management infrastructure especially in the remaining critical woodland habitat. Sometimes loge management and staff do not live in respect and compatible with the biodiversity conservation objectives of the park.

Table 7. The area covered by administrative and tourism facilities inside Amboseli

Structure for tourism or administration	Area (km^2)	Proportion of area (%) in Amboseli National Park	Remarks about location of the administrative and tourism facilities
Roads network (total length of 250 km)	1.02	0.26	The road network traverse throughout the park with major roads cutting right across it
Gates and ranger accommodation area: Kimana Merito Kitirua Eremito	0.0441265	0.01485	There are four main gates to Amboseli National Park. Ranger houses for accommodation accompany these gates inside the park. The four gates are about equidistantly placed
Park administration block and ranger accommodation	0.251607014	0.06419	The park headquarters and associated ranger living units is placed inside the park and in the rare and declining acacia woodland which also supports the highest large species diversity in Amboseli
Lodges areas fenced off by electric fence: Serena Amboseli Oltukai	3.0776	0.70	There are two main lodge locations in Amboseli National Park. Serena Hotel occupies its own location near a second major swamp. Three other hotels (Oltukai, Amboseli and one collapsed one) are located near the biggest swamp almost at the center of the park and fenced by electric fence and with many living units, tourists and workers.
Airstrip with administration block	0.0244	0.0836	Located near the Oltukai – Amboseli main lodge area and has an administrative building
Two electrically fenced enclosures	0.7653938	0.19525	Exclosures for scientific study to show that vegetation regeneration can occur in the absence of elephants. These have become permanent and taken wildlife space in Amboseli National park
Elephant research headquarter	0.00882	0.00225	A tiny administrative area for the long – lasting and world famous Cynthia Moss" African Elephant Research Project.
Observation Hill with toilets	0.080384	0.020506	This area is a small raised hill in Amboseli from which visitors can see the full view of the park and Mt. Kilimanjaro. It is supplied with toilets and visitors move out of their vehicles. Wildlife rarely venture here during the day because of human presence
TOTAL	5.27	1.43%	

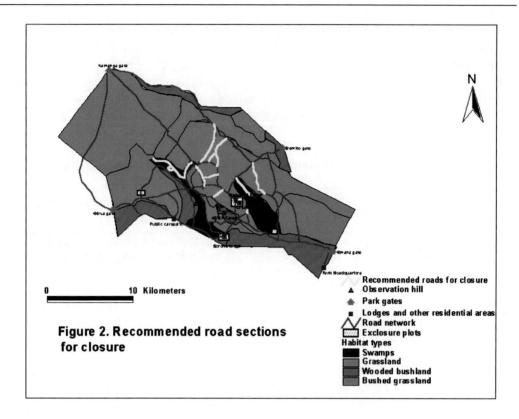

Figure 2. Recommended road sections for closure

Figure 1. The road network in Amboseli National Park. There are more roads than necessary and closing some (highlighted in yellow) will not still not compromise tourism viewing in the park. The roads highly fragment this park in all directions, making it difficulty to balance tourism and conservation of biodiversity in the park.

The road network (figure 1) is relatively high for the park of this size. Many dry season roads have arisen from illegal off – road driving activities. Such roads fragment and degrade the park more. The road network creates edge effect and destroys the habitats. With over – speeding of vehicles, they become a source of animal harassment and death. To reduce the fragmentation effect, we think that visitors will still see all common animals in Amboseli (which is relatively open) with a reduction of up to 30% of the current road network in Amboseli (figure 2). The park should also consider purchasing land immediately outside the park to relocate the airstrip and establish all ranger accommodation because the effects of these facilities consume conservation space and impact negatively on the park. We recently observed further expansion of ranger accommodation facilities at the park headquarters within the diminishing woodland habitat. This should be discouraged and alternative accommodation outside the park sought.

The area controlled by Olkajiado County Council in which Ol Tukai and Amboseli Lodge are located is a big threat to conservation objectives of Amboseli. The number of litter that attracts baboons and unplanned human structures in servant quarters and sheer large human traffic at this location almost negates conservation gains in Amboseli. Negotiation with the county council to have strong environmental requirements and human activities at this location is critical as it is located right at the heart of Amboseli and its negative activities diffuse and compromise Amboseli as a conservation area.

Conclusion

Amboseli National Park is a critical asset to Kenya in terms of tourism potential, tourism achievement and generation of revenue. It is also an assets as a protected area for biodiversity conservation. However, it is faced by hosts of threats both inside and outside the park. The high elephant density and associated destruction of woodland has influenced species density and diversity by creating dominance of plain grazers inside the park. Further, a high road network, ranger accommodation facilties and tourism infrastructure sometimes located in sensitive swamp and woodland habitats, threatens the park from within. To the outside, the park has been insularized by changing land uses (especially agriculture), impending group ranch sub – division, electric fences for irrigation schemes (such as Kimana and Namelok) that are blocking the corridor and dispersal area, expanding markets centers and increasing human encroachment and settlement that is diminishing the effective wildlife dispersal and migratory area outside the park. The management has to think of a proper elephant management strategy for the future, proper placement and management of tourism facilities and those for administration, and have to address how Amboseli will survive in a changing landscape. To have Amboseli to continue to generate tourism revenue needs re- strategizing to reduce threats both within the park and to its dispersal area outside to balance both biodiversity, tourism interests, and incorporate local human dimension in the viability of Amboseli National Park.

References

Berger, D.J. (1993). *Wildlife Extension*. Participatory Conservation by the Maasai of Kenya. African Center for Technology Studies Nairobi, Kenya. Pg. 25.
Beresford, M. & Philips, A. (2000). Protected Landscapes: A Conservation Model for the 21st Century. Forum: *Journal of the George Wright Society*. **17**(1): 35-54.
Bourn, D. & Blench, R. (eds.) (1999) *Can Livestock and Wildlife Coexist?: An interdisciplinary approach*. Overseas Development Research Study. 251 pp.
Burkey, T. (1994) "Faunal Collapse in East African Game Reserves Revisited." *Biological Conservation*. **7**(1): 107-110.
Campbell, D.J., Gichohi, H, Mwangi, A, & Chege, L. (2000). Land use conflict in Kajiado district, Kenya. *Land Use Policy* **17**:337 – 438.
Fratkin, E. (1994). Pastoral Land tenure in Kenya: Maasai, Samburu, Boran, and Rendille experiences, 1950-1990. *Nomadic Peoples* **34**:55-68.
Gichohi, H., Gakuhu, C & Mwangi, E. (1996). Savannah Ecosystems. Pages 273-298 In. *East African Ecosystems and their Conservation*. McClanahan, T. R., & Young, T.P. (eds.), Oxford University Press, Oxford.
Honey, M. S. (1999). Treading Lightly?: ecotourism's impact on the environment. *Environmen*. **41**(5): 4-18.
Kenya Wildlife Service (2001). Amboseli too Small for Its Wildlife. KWS. News Issues No.2
MacKinnon, J., Child, K., & Thorsell, J. *(eds.)* (1986) Managing Protected Areas in the Tropics. International Union for the Conservation of Nature and Natural Resources, Gland, Switzerland. 295pp.
Mwale, S. (2000) Changing Relationships. *Swara* **22**(4):11-17.

Newmark, W. D. 1996. Insularization of Tanzanian parks and the local extinction of large mammals. *Conservation Biology*. **10**(6): 1549-1556.

Newmark, W. D. & Hough, J. L. (2000). Conserving Wildlife in Africa: Integrated Conservation and Development Projects and Beyond. *Bioscience* **50**(7): 585-592

Okello, M.M. & Kiringe, J.W. (2004). Threats to Biodiversity and the Implications in Protected and adjacent dispersal areas of Kenya. *Journal for Sustainable Tourism*. **12**(1): 55 – 69

Okello, M. M. (2005). Land Use Changes and Human - Wildlife Conflicts in the Amboseli Area, Kenya. *Human Dimensions of Wildlife* **10**(1): 19 – 28

Okello, M. M., B.E.L. Wishitemi, & B. Lagat. (2005). Tourism Potential and achievement of Protected Areas in Kenya: Criteria and Prioritization. *Tourism Analysis* **10** (2):151 – 164

Okello, M. M., Wishitemi, B. E, & Mwinzi, A.M. (2001). Relative Importance of Conservation Areas in Kenya Based on Diverse Tourist Attractions. *The Journal of Tourism Studies*. **12**(1):39 - 49

Okello, M. M., Seno, S.K., & Wishitemi, B. E. L. (2003) Maasai community wildlife sanctuaries in Tsavo – Amboseli Ecosystem, Kenya: management partnerships and their conditions for success. *The Parks Journal* 13(1). IUCN Gland, Switzerland.

Sarkar, S. (1999). Wilderness preservation and biodiversity conservation: Keeping divergent goals distinct. *Bioscience* **49** (5): 405 – 411.

Seno, S. & Shaw W.W. (2002). Land tenure policies, Maasai traditions and wildlife conservation in Kenya. *Society and Natural Resources*. **15**: 79-88.

Sindiga, I. (1995). Wildlife-based tourism in Kenya: Land use Conflicts and government compensation policies over protected areas. *The Journal of Tourism Studies* 6(2):45-55.

Soule', M. E., Wilcox, B. A Holtby C. (1979). Benign neglect: a model of faunal collapse in game reserves of East Africa. *Biological Conservation* **15**: 259 – 272.

Western, D. (2002). In the dust of Kilimanjaro. Island Press / A shear Water Book. Washington, DC. 309 pp.

Western, D. (1989). The Ecological role of Elephants in Africa. *Pachyderm*, **12**, 42-45

Western, D. (1992). Conserving *Savanna Ecosystems Through Community Participation: The Amboseli Case Study*. Prepared for the Liz Claiborne Art Ortenberg Foundation Community Based Conservation Workshop. 71pp.

Western, D. (1997) Nairobi National Park is Slowly Being Strangled By Development. *Swara* **19**(6) and 20(1):19-20.

Western, D., & Ssemakula, J. (1981). The future of savanna ecosystems: ecological islands or faunal enclaves? *African Journal of Ecology* **19**: 7 – 19.

White, I. J. (2001). Headaches and heartaches: The elephant Management Dilemma. In: Environmental Ethics: Introductory readings. Eds Schmidtz, D. & Willot, E. pp 293 - 305. New York: Oxford University Press.

Whittaker, R. (1998) *Island Biogeography: Ecology, Evolution, and Conservation*. Oxford University Press, New York. 33-193pp.

Wishitemi, B.E.L. & Okello, M. M. (2003) Application of the Protected Landscape Model in Maasai communally owned lands of southern Kenya. *The Parks Journal* **13**(2): In Press. IUCN Gland, Switzerland.

Young, T.P. & McClanahan, T.R. (1996) Island Biogeography and Species Extinctions. pp292-293. In: *The East - African Ecosystems and their Conservation*. McClanahan, T.R. and T.P. Young (eds.). Oxford University Press, Oxford.

In: National Parks: Biodiversity, Conservation and Tourism
Editors: A. O'Reilly and D. Murphy, pp. 131-146
ISBN: 978-1-60741-465-0
© 2010 Nova Science Publishers, Inc.

Chapter 7

SACRED GROVES: INFORMAL PROTECTED AREAS IN THE HIGH ALTITUDES OF EASTERN HIMALAYA, ARUNACHAL PRADESH, NORTHEAST INDIA: TRADITIONAL BELIEFS, BIODIVERSITY AND CONSERVATION

A.R. Barbhuiya[*a], *M.L. Khan*[a], *A. Arunachalam*[a], *S.D. Prabhu*[b] *and V. Chavan*[b]

[a] Department of Forestry, North Eastern Regional Institute of Science & Technology, Arunachal Pradesh, India
[b] Digital Information Resource Laboratory, National Chemical Laboratory, Pune, India

Abstract

Sacred groves are 'traditionally managed' forest patches that functionally link the social life and forest management system of a region. They are the repositories of economical, medicinal, rare, threatened and endemic species and can be regarded as the remnant of the primary forests left untouched/undisturbed by the local inhabitants and protected by local communities due to beliefs that the deities reside in these forests. Arunachal Pradesh, the 'land of rising sun', is located in the northeast region of India, sharing international boundaries with Bhutan, China, Tibet and Myanmar, and is unparalleled in the world for its concentrations, isolation and diversity of tribal cultures and biological diversity. Thus it falls under one of the eight global *mega-diversity hotspots* in the world. It lies between $91°30'$–$97°30'$E longitudes and $26°28'$–$29°30'$N latitudes, covering an area of 83,743 km^2. Approximately 94% of the area is covered by forests, 17.21% of which is very dense, and 45.35% moderately dense. Open areas comprise 18.38% and nonforest areas comprise 18.91%. With a tribal population of about one million represented by 21 major tribal groups with more than 100 ethnically distinct subgroups and over 50 distinct dialects, Arunachal Pradesh contains a good number of

---

* E-mail address: arbarbhuiya@gmail.com. A.R. Barbhuiya, Department of Forestry, Mizoram University, Aizawl-796009, Tanhril, Mizoram, India (Correspondence author).

sacred groves particularly attached to the Buddhist monasteries, called *gompa forests*, which are managed by the Buddhist community (Monpa and Sherdukpens) of Arunachal Pradesh. These monasteries are mainly found in the West Kameng and Tawang districts of Arunachal Pradesh. Besides these gompa forests, there are a good number of sacred groves in the Tawang and West Kameng districts of Arunachal Pradesh related to the community culture and beliefs. So far, about 63 such sacred groves have been explored as part of a pilot study, including the geographic information, physical and biological attributes and traditional myths associated with the sacred grove. Besides a number of formal protected areas in the region, these informal protected areas, i.e., sacred groves, also play a vital role in the conservation of the significant biodiversity of the region. In the study area, which is more highly valued than the other parts of Arunachal Pradesh, the local Buddhist community also provides tourism services to visitors. The tourism potential in relation to sacred groves and cultural resources is great, but is lacking at present, and needs to be evaluated and appreciated with the active participation of local communities, government bodies and NGOs for the better and sustainable biodiversity services as well as ecotourism. Thus, the rich and enormously diverse biodiversity and cultural heritage of the region may also provide an opportunity for the development of culture tourism in the region. In this regard, the numerous tribes and their multifaceted fairs and festivals can be a powerful attraction to magnetize tourists. An existing scenario of tribal transformation has, however, put pressure on these sacred groves of different sizes. Thus, pro-conservation activities are also admissible for biodiversity management in the Indian eastern Himalayan region and Arunachal Pradesh, northeast India.

Key Words: Arunachal Pradesh, Buddhism, biodiversity; gompa, Monpa, sacred grove

Introduction

Sacred groves are traditionally managed forest patches that functionally link the social life and forest management system of a region. The tracts of virgin forest harboring rich biodiversity, protected by the local people based on indigenous and religious beliefs and taboos are called sacred groves. They are the repositories of economical, medicinal, rare, threatened and endemic species and can be regarded as the remnant of the primary forests left untouched/undisturbed by the local inhabitants and protected by local communities due to beliefs that the deities resides in these forests. Many workers have described these sacred groves in different ways, and most of them have emphasized the natural or near-natural state of vegetation in the sacred groves and the preservation of these groves by local communities through social taboos and sanctions that reflect the spiritual and ecological ethos of these communities (Malhotra et al., 2001; Ramakrishnan, 1998). The role of sacred groves in the conservation of biodiversity has long been recognized (Kosambi, 1962; Gadgil and Vartak, 1975; Khan et al., 1997; Khumbongmayum et al., 2005).

Sacred groves are found all over India, especially in those regions inhabited by indigenous communities. The existence of sacred groves all along the Himalayas from northwest to northeast, the central Himalayas of Kumaon and Garhwal, Darjeeling and Meghalaya has been reported by Roy Burman (1992). Northeast India, with its varied physiography, soil and climate, supports different types of forests such as tropical, subtropical, temperate and alpine forests. Many natural disturbances such as landslides, cyclones, floods, and forest fires, and anthropogenic disturbances like shifting cultivation, illegal felling of trees, and NTFP (non-timber forest products) collections have led to the destruction of many primary forests and ultimately the development of secondary forests. However, despite these anthropogenic disturbances, many forest patches are still covered by the sacred groves, which are being managed and protected by the tribal communities based on religious beliefs. In spite of rapid modernization, traditional ecological beliefs continue to survive in many local and non-tribal societies of northeast India, although often in reduced form, and many of them are in different stages of degradation. Most of the sacred groves are located near human settlements, so anthropogenic disturbances in these forests are progressively increasing. Many indigenous ethnic groups/tribes inhabit northeast India and, according to their traditional concepts, gods/deities reside in many forest areas. The management of such sacred groves through informal norms, ethical rules and religious beliefs has resulted in thousands of such sacred forests.

Figure 1. Map of West Kameng and Tawang districts of Arunachal Pradesh.

Arunachal Pradesh in the Indo-Burma biodiversity hotspot is unparalleled in the world at present for its concentrations, isolation and diversity of tribal cultures and biological diversity. With a population of about one million, it is a 70% tribal state, containing 21 major tribal groups with more than 100 ethnically distinct subgroups and more than 50 distinct dialects. In Arunachal Pradesh, there are a few sacred groves attached to the Buddhist monasteries, called Gompa Forest Areas (GFAs), which are managed by Lamas and Monpa tribes of Arunachal Pradesh. These monasteries are mainly found in the West Kameng and Tawang districts of Arunachal Pradesh. Malhotra et al. (2001) reported that in Arunachal Pradesh there are 58 GFAs and other districts, namely Lower Subansiri and Siang, which also have sacred grove forests (Chatterjee et al., 2000). Nonetheless, information on this aspect is scattered at various levels without proper documentation. With the rich cultural and biological richness of the region, it is evident that the human tradition is in nexus with natural resources, particularly the forest. Thus, the present inventory holds significance, and the study documents physical, socio-cultural and biological status of sacred groves in the western part of Arunachal Pradesh that is mostly upland.

Table 1. Physiographic profile of Tawang and West Kameng districts of Arunachal Pradesh

District	Area (km^2)	Population* (%)	Literacy rate*	Forest Cover (km^2) Evergreen	Degraded & Deciduous	Forest Type
Tawang	2172	39242	41.14	870	582	Temperate broad leaved forests
	(2.59)	(3.57)		(1.68)	(1.12)	and temperate conifer forests
West Kameng	7422	74527	61.67	3571	1411	Subtropical and pine forests
	(8.86)	(6.78)		(6.93)	(2.73)	

Values in parentheses are the % of the respective total values of the state as a whole
Source: Statistical Abstract of Arunachal Pradesh; *as per 2001 Census (Anonymous 2004)

Materials and Methods

Study Area

Arunachal Pradesh, the 'land of rising sun', formerly known as NEFA (North Eastern Frontier Agency), is situated in the extreme northeastern region of India. It extends between 91°30′–97°30′E longitudes and 26°28′–29°30′N latitudes covering an area of 83,743 km^2. The study was conducted in West Kameng and Tawang districts of Arunachal Pradesh (Figure 1). The West Kameng is located in the western part of Arunachal Pradesh covering an area of 7,422 km^2 (Table 1). It is bounded by China on the north, East Kameng district on the east and Sonitpur district of Assam on the south and on the west by Bhutan and Tawang district of Arunachal Pradesh itself. It lies in the 26°54′–8°01′N latitude and 91°30′–92°40′E longitude. The district is mostly mountainous and a greater part of it falls within the higher mountain zone consisting of mass tangled peak and valleys. The main river of the district is Kameng River with lots of seasonal and perennial streams and streamlets. The Tawang district covering an area of 2172 km^2 and lies between 27°25′–27°52′N latitude and 91°16′–91°59′E longitude. It is surrounded by Tibet in the north, Bhutan in the southwest and Sela ranges separate from West Kameng district in the east. Approximately two-thirds of the entire area of the district is highly mountainous. There are two main rivers, viz., Tawang-Chu and Nyamjang-Chu. Tawang and West Kameng districts land use pattern and tenancy is based on the customary and traditional system of the state.

The local people exercises traditional right over land which is again held individually, commonly and clan wise basis for agricultural purposes like jhuming (burning), settled, terrace and wet rice cultivation. They also exercise right over land for traditional hunting, fishing, grazing and extraction of forest products for domestic purposes. The predominant vegetation in some of the areas of West Kameng district is subtropical broad leaved and pine forests types of evergreen and dense in nature (Table 1). A few areas of West Kameng, i.e., Bomdila and Dirang and Tawang district of Arunachal Pradesh exhibit temperate broadleaved and temperate coniferous forests. The vegetation is open and is not stratified. However, the forest was dominated by oaks and members of Magnoliaceae and Ericaceae, and particularly the Rhododendrons.

Survey Methods

An extensive filed survey was undertaken to inventoried the sacred groves of Arunachal Pradesh during 2005–2006. Records of the state government and literature were consulted to locate the groves and to ascertain their historical backgrounds. Traditional institutions administered by Gaon-bura (village headmen), Lama (priests of Buddhist monasteries) and local people, educated persons, caretakers of the sacred groves were contacted for identifying sacred forests in the territories under their control or in their knowledge. The physical (physical location was recorded by using GPS), socio-cultural and biological features associated with each sacred forest was collected through a questionnaire, followed by field visits and interactions with elderly people and prominent citizens to asses the attitudinal changes and their suggestion for conservation. The species composition was studied in all the visited sacred forests through visual observations and the species were identified by consulting different monographs, viz., *The Flora of British India*, *The Flora of Assam*, *Materials for the Flora of Arunachal Pradesh* and consultations with floral herbarium of State Forest Research Institute, Itanagar and Botanical Survey of India (Itanagar Branch).

Results and Discussion

A total of 63 sacred groves were inventoried with detailed information from the two districts of Arunachal Pradesh during the study period (Table 2). Maximum numbers (39) of these are located in the Tawang district and minimum in the West Kameng (24) district of Arunachal Pradesh. Among the 63 sacred groves, 38 have been found adjacent to the Gompa, i.e., Buddhist monasteries, and they are under the control of monasteries and conserved by religious faith. Another 25 groves in the area have been conserved from spiritual point of view or by fear of specific deities and according to their belief that these forest patches are the property of evils spirit/gods/deities and must therefore not be damaged in any way (Table 2).

Table 2. Physiographic details of the sacred groves located in the high altitudes of Arunachal Pradesh

Sl. No.	Name of the sacred grove	Latitude (N) (range)	Longitude (E) (range)	Altitude (mean) (m)	Terrain type	Area (km^2)	DT.	BL.	Nearest settlement
1	Arkidung	27.34.14-27.34.58	91.54.07-91.54.31	2463.00	HS	0.25	T	T	Lhou Village
2	Bomdila Bazar Line	27.15.00-27.15.58	92.24.11-92.24.48	2637.00	HS	0.01	W. K	B	Bomdila Bazar
3	Brakar	27.34.15-27.34.48	91.56.11-91.56.48	2975.62	HS	0.30	T	T	Lhou Village
4	Chambu	27.35.18-27.35.44	92.52.13-92.52.38	1786.35	HS	0.50	T	T	Chambu Village
5	Dungarmani	27.35.09-27.35.49	92.00.18-92.00.56	2576.90	HS	0.50	T	J	Namazing Basti
6	Dungirmoon	27.34.30-27.34.56	91.58.22-91.58.49	2380.40	HS	0.30	T	J	New Kharsa Basti
7	Dupphang	27.19.01-29.19.34	92.16.33-92.16.55	1941.56	HS	0.25	W. K	D	Dirang Basti
8	G. G. Rabgyeling[1]	27.16.07-27.16.48	92.25.08-92.25.31	2638.50	HT	0.50	W. K	B	Bomdila Bazar
9	G. G. Rabgyeling[2]	27.18.13-27.18.51	92.26.07-92.26.33	2702.25	HT	1.10	W. K	B	Bomdila Bazar
10	G. Naymgal Lhatse	27.35.07-27.35.57	91.51.27-91.52.26	2911.52	HT	1.50	T	T	Sheyo Basti
11	Gazangphrang	27.26.00-27.26.41	92.18.13-92.18.54	1958.45	HS	0.40	W. K	D	Dirang Basti
12	Gomluk	27.20.34-27.20.55	92.16.08-92.16.22	1952.33	HS	0.70	W. K	D	Dirang Basti
13	Grang	27.23.09-27.23.48	92.11.03-92.11.50	1684.00	HS	2.60	W. K	D	Yang Village
14	Jung Bara Basti[1]	27.34.52-27.34.55	91.58.55-91.58.58	2366.35	HS	0.01	T	J	Jung Bara Basti
15	Jung Bara Basti[2]	27.34.08-27.34.55	91.59.03-91.59.44	2754.60	HS	1.50	T	J	Jung Bara Basti
16	Jung Zero Point	27.34.13-27.34.39	91.59.03-91.59.33	2408.50	HS	1.20	T	J	Jung Zero Point
17	Jyoti Nagar	27.21.11-27.21.54	92.14.12-92.14.30	1832.45	HS	0.50	W. K	D	Jyoti Nagar
18	Kakaling	27.35.00-27.35.11	91.52.01-91.52.15	2701.55	HT	0.30	T	T	Tawang
19	Kalchakra	27.21.09-27.21.55	92.14.08-92.14.34	1649.56	V	1.20	W. K	D	Jyoti Nagar
20	Kepsingmu	27.19.01-27.19.58	92.13.08-92.14.37	2137.00	HS	3.00	W. K	D	Saksem Basti
21	Khadun	27.20.35-27.20.48	92.16.09-92.16.22	1615.00	HT	0.75	W. K	D	Dirang Basti
22	Kharsa Basti[1]	27.34.44-27.34.45	91.58.19-91.52.20	2237.56	HS	0.01	T	J	Kharsa Basti
23	Kharsa Basti[2]	27.34.00-27.34.49	91.50.16-91.50.44	2488.65	HS	0.50	T	J	Kharsa Basti
24	Khartang	27.34.22-7.34.524	91.50.19-91.50.46	2468.33	HS	0.60	T	J	Khartung Basti

Table 2. Continued.

Sl. No.	Name of the sacred grove	Latitude (N) (range)	Longitude (E) (range)	Altitude (mean) (m)	Terrain type	Area (km2)	DT.	BL.	Nearest settlement
25	Khartung Basti	27.34.31-27.34.37	91.58.16-91.58.16	2341.00	HS	0.01	T	J	Khartung Basti
26	Khartung	27.35.09-27.35.23	91.55.14-91.55.38	2896.00	HS	1.10	T	T	Khirmu Village
27	Khinmey	27.35.08-27.35.20	91.53.0991.53.44	2961.25	HS	0.50	T	T	Khinmey Village
28	Khirmu	27.34.14-27.34.55	91.55.32-91.55.56	2974.33	HS	1.50	T	T	Khirmu Village
29	Khraling Basti	27.34.09-27.34.11	91.58.59-91.58.59	2352.00	HS	0.01	T	J	Khraling Basti
30	Kulangaus Basti	27.34.50-27.34.51	91.58.33-91.58.36	2314.45	HS	0.01	T	J	Kulangaus Basti
31	Labrang	27.35.08-27.35.19	91.52.08-91.52.30	2955.44	HS	0.70	T	T	Lebrang Basti
32	Lieung	27.20.00-27.20.53	92.15.02-92.15.40	1855.72	HS	1.20	W. K	D	Lieung Basti
33	Lower Shera Basti	27.17.08-27.17.48	92.26.12-92.26.40	2436.48	HS	1.60	W. K	B	Lower Shera Basti
34	Maishing	27.18.13-27.18.56	92.14.00-92.14.51	1845.52	HS	0.50	W. K	D	Maishing Basti
35	Mani	27.21.04-27.21.48	92.15.03-92.15.36	1855.63	HS	1.50	W. K	D	Dum Dirang Basti
36	Manidungur	27.21.04-27.21.49	92.14.04-92.14.33	1792.65	HS	0.30	W. K	D	Yang Village
37	Namazing Basti	27.34.53-27.34.55	91.58.28-91.58.30	2252.00	HS	0.01	T	J	Namazing Basti
38	Namet	27.34.08-27.34.49	91.54.09-91.54.38	3011.08	HS	0.50	T	T	Namet Basti
39	Neharu	27.35.08-27.35.35	91.52.33-91.52.39	1688.67	HT	0.20	T	T	Neharu Colony
40	New Bomdila	27.09.24-27.09.33	92.23.41-92.24.06	2041.05	HT	0.03	W. K	B	New Bomdila Basti
41	New Kharsa Basti	27.33.40-27.33.42	91.58.55-91.58.56	2577.67	HS	0.01	T	J	New Kharsa Basti
42	Nuranang	27.34.55-27.34.57	91.58.58-91.58.59	2350.40	HS	0.01	T	J	Jung Bara Basti
43	P. J. Dhargyelling	27.20.18-27.20.52	92.15.07-92.15.27	1723.50	HT	3.00	W. K	D	Lieung Basti
44	P. New Lebrang	27.35.00-27.35.20	91.52.11-91.52.29	2880.25	HS	0.30	T	T	New Lebrang Basti
45	Regilling[1]	27.95.00-27.95.48	91.52.09-91.52.46	2945.80	HS	0.13	T	T	Damnian Basti
46	Regilling[2]	27.34.38-27.35.03	91.52.02-91.52.48	2800.30	HS	3.00	T	T	Urgelling Basti
47	Rimpopha	27.19.04-27.19.33	92.14.30-92.14.56	1586.00	HS	0.50	W. K	D	Dirang Basti
48	Sangdok Palri	27.34.18-27.34.39	91.58.10-91.58.44	2560.35	HT	1.00	T	J	Sangdakpalri Basti

Table 2. Continued.

Sl.No.	Name of the sacred grove	Latitude (N) (range)	Longitude (E) (range)	Altitude (mean) (m)	Terrain type	Area (km2)	DT.	BL.	Nearest settlement
49	Sangechulen	27.21.05-27.21.35	92.14.00-92.14.48	1872.40	HS	1.50	W. K	D	Youang Basti
50	Sharmang	27.33.14-27.33.48	91.55.09-91.55.34	2948.16	HS	0.30	T	T	Namet Basti
51	Sheyo	27.35.04-27.35.33	91.51.08-91.51.44	3118.95	HS	1.00	T	T	Sheyo Basti
52	Singsur Ani	27.33.00-27.33.31	91.54.08-91.54.25	2846.67	HS	0.50	T	T	Namet Basti
53	Sunglingnang	27.19.11-27.19.46	92.14.12-92.14.54	1985.20	HS	0.55	W. K	D	Burchi Basti
54	Thespa	27.34.00-27.34.21	91.54.07-91.54.48	2698.40	HS	0.50	T	T	Namet Basti
55	Thongmen	27.35.08-27.35.29	91.55.05-91.55.38	3172.88	HT	0.40	T	T	Khirmu Village
56	Tsangpu	27.33.08-27.35.53	91.52.08-91.52.51	2722.00	HT	0.11	T	T	Tsangpu village
57	Upper Shera Basti	27.13.40-27.13.53	92.26.07-92.26.49	2684.32	HT	0.80	W. K	B	Shera Basti
58	Urgilling	27.34.00-27.39.52	91.52.04-91.52.55	2851.80	HT	0.10	T	T	Urgelling Basti
59	Yang	27.36.00-27.3652	91.52.00-91.52.33	2957.45	HS	3.00	T	T	Chambu Village
60	Yangfaundi	27.19.11-27.19.56	92.14.04-92.14.58	1653.34	HS	2.00	W. K	D	Richong Basti
61	Yidgachoszin	27.35.11-27.35.30	91.52.08-91.52.19	2684.00	HT	0.50	T	T	Lebrang Basti
62	Youngmang	27.35.09-27.35.44	91.52.22-92.52.56	2988.50	HS	0.75	T	T	Chambu Basti
63	Youngnang	27.21.04-27.21.32	92.14.03-92.14.11	1746.30	HT	0.50	W. K	D	Young Basti

DT- District: T-Tawang; W.K.-West Kameng;
BL-Block: T-Tawang; B-Bomdila; D- Dirang; J- Jung
HS-Hill slope; HT- Hill top, V- Valley. Dist. to S.C- Distance to State Capital.

Most of the sacred groves have well demarcated boundaries with a few exceptions. The size of the individual sacred groves varied from a clump of a few trees having an area of 0.01–3.00 km^2 (Figure 4) with the elevation of 1615–3200 m and most of them are located in the hill-slope followed by hill-top terrain (Figure 2 a). The area of the sacred groves smaller in size compared to those in the other states of Northeast India like Meghalaya (Tiwari et al., 1999), which varied between 0.01–9.00 km^2. However, the present records are similar to that of Manipur (0.01–4.00 km^2) as reported by Khumbonmayum et al. (2004) and Khumbonmayum et al. (2005).

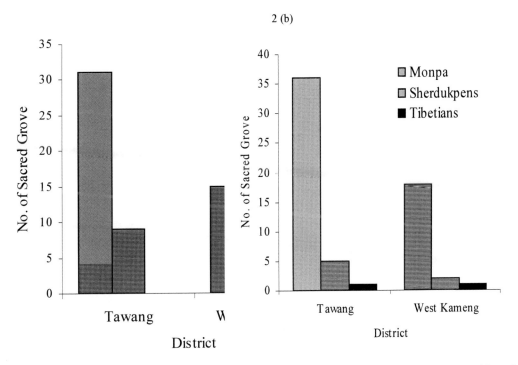

Figure 2. Distribution of sacred groves in (a) different terrain types and among the three communities of Tawang and West districts of Arunachal Pradesh.

Brief History and Socio-cultural Beliefs of these Sacred Groves

The practice of keeping sacred groves was prevalent among the different tribes depending on their cultural practices. Almost all inhabitants of the study area are Monpas, along with a few Sherdukpens and Tibetan tribes (Figure 2 b). The latter call themselves Monpas who were, however, settled in Tawang and West Kameng districts of Arunachal Pradesh after Chinese aggression in 1962 in Tibet. The Monpas follow Mahayana Buddhism and their religious activities take place around the gompas (monasteries) located in the different parts of the state and most of them are in the high altitudes. All of these gompas are considered sacred by the Monpas and hence all the objects including plants and animals in the gompa area are protected as sacred assets of the gompa. Before the advent of Buddhism, Monpas believed in the 'Bon' faith, which was characterized by spirit worship and animal sacrifice. Many gods, spirits and demons held to represent the powers and forces of their wild landscape and were believed to be responsible for all the calamities threatening their welfare. So they had to take the assistance of the priest to protect themselves from their bad activities as well as to assure their own welfare. When the Monpas adopted Buddhism, it stopped these rites and accepted these indigenous supernatural being its pantheon as the local defender of its faith and in return for such services, they become entitled to be worshipped according to the manner of the Buddhist rituals. A few of these supernatural beings, found unsatisfied with the new rituals, are again being worshipped according to the old rituals and with the help of religious functionaries of the old religion. One of such beliefs is the worship of a spirit called *Naga* (*Luhu*) who, people say, lives in some stone structures, forests or caves near each household or in some places in the village. This spirit is believed to control disease and other family problems. Whenever some problems come in the society they sacrifice rituals near these identified structures to satisfy the spirit for the welfare of the family or the community.

Each of the Monpa villages has its own distinct territorial boundary and demarcated either by some streams, stone pillar, stone mark or by a forest termed *santsum*. Such forests are preserved as they cherish the belief that such trees are the abode of some serpent spirit or demons termed *Luhu* and *Zipda* and therefore they refrain from felling the trees to avoid effects of these spirits (Lama 1999). Monpas usually perform many sacrifice rituals to please their deities/spirits for which they use particular plants that considered sacred. These species include *Juniperus squamata* (Shukpa), *Rhododendron formasum* (Sulu), *R. anthopogon* (Ballu), *Cryptomaria japonica*, *Lycopodium sp.* (Thimpo), *Nardostachys jatamansii* (Spang spos), etc.

Vegetation in Sacred Groves

Sacred groves are found to be the best centre for the conservation of plant diversity. Most of the sacred groves may be considered the remnants of the climax evergreen forests natural to the region. Nevertheless, species composition of vegetation in the sacred groves varied with the difference in topography vis-a-vis forest type. Maximum number (109) of plant species was recorded in the sacred groves of West Kameng district than in the Tawang district of Arunachal Pradesh due to extreme agro-climatic regime in the latter. The contribution of tree species was more in the sacred groves of both the districts followed by herbs and shrubs (Figure 3 a). Maximum number of rare plant species (13) was also recorded in the sacred groves of West Kameng district and there were 11 in the sacred groves of Tawang (Figure 3 b). Again, this could be attributed to varied environmental factors like physiography, geology, soil, climate, etc. The majority of the tree species are subtropical evergreen, with a few deciduous species in the West Kameng district of Arunachal Pradesh. In Tawang most of the species are temperate evergreen along with deciduous species (Table 3). Over all, the trees of the sacred groves are large in size, reaching heights of over 30 m.

The sacred groves in Arunachal Pradesh are important for the conservation of endemic and endangered species. For instance, species like *Aconitum spictum, Cinnamomum tamala, Juglans regia, Mesua ferrea, Rhododendron arboreum, Swertia chirata, Thalictrum foliolosum, Taxus wallichiana, Zanthoxylum oxyphyllum* that have been listed as threatened plants of Darjeeling Himalaya by Chhetri et al. (2005) were present in the sacred groves recorded in Arunachal Pradesh. Other commercially and medicinally important, but ecologically threatened species such as *Aconitum ferox, Cimmamomum tamala,*

Cimmamomum zeylanica, Swertia chirata, Taxus wallichiana, etc., were also found in the different sacred groves in the Tawang and West Kameng districts of Arunachal Pradesh. Presence of primitive angiosperms in the sacred groves such as *Alnus nepalensis, Magnolia graffithi, Castonopsis indica* and other species of *Michelia* and *Rhododendron* suggest that the sacred groves are also a repository of evolutionarily important plant species. Takhtajan (1969) has suggested the northeastern region of India as the "Cradle of Flowering Plants" due to the presence of a large number of primitive angiosperms.

Table 3. List of trees, shrubs and herbs in the sacred groves of high altitudes of Arunachal Pradesh

Species	Habit	Species	Habit
Acer palmata	Tree	Malus domestica	Tree
Acer pectinatum	Tree	Melocana baccifera	Herb
Acogonum mole	Herb	Mesua ferrea	Tree
Aconitum ferox	Herb	Michelia champaca	Tree
Aconitum spictum	Shrub	Mikenia micrantha	Herb
Alnus nepalensis	Tree	Mimosa pudica	Herb
Anacardium oxidentalis	Tree	Mimusops elengi	Shrub
Anaphalis triplinervis	Herb	Myrraya exotica	Tree
Anemone elogata	Herb	Myrsine semiserrata	Tree
Anthocephalus kadamba	Tree	Oxalis corniculata	Herb
Argemone mexicana	Herb	Oxalis martiana	Herb
Australian acacia	Tree	Panax bipninatifibus	Shrub
Azadirachta indica	Tree	Persea bombycina	Tree
Bambusa balcooa	Herb	Phlogonthes sp.	Herb
Bambusa tulda	Herb	Phyllanthes embalica	Tree
Berberis aristata	Shrub	Phyllostachys bambusoides	Herb
Berberis vulgaris	Shrub	Picrorhiza kurooa	Shrub
Betula utilis	Shrub	Pinus kesiya	Tree
Blumea glomerata	Herb	Pinus roxburghii	Tree
Borreria hispida	Herb	Pinus wallichiana	Tree
Bulbophyllum sp.	Shrub	Podophyllum hexandrum	Herb
Caesalpinia pulchirema	Tree	Polyalthia longifilia	Tree
Callistemon lanceolatus	Tree	Polygonum glabrum	Herb
Camelia acuminate	Tree	Polygonum multiflorum	Herb
Camelia caudata	Tree	Prunus domestica	Tree
Cannabis sativa	Shrub	Puncea granatum	Tree
Cassia fistul	Tree	Pyrus mallus	Tree
Castonopsis indica	Tree	Pyrus pashia	Shrub
Cinnamomum tamala	Tree	Quercus griffithi	Tree
Cinnamomum zeylanica	Tree	Quercus incana	Tree
Citrus reticulata	Tree	Quercus kamroopii	Tree
Clerodendron infortunatum	Shrub	Quercus lamellose	Tree
C. viscosum	Shrub	Quercus lanata	Tree
Clestemon lanceolatus	Tree	Quercus semicarpifolia	Tree

Table 3. Continued.

Species	Habit	Species	Habit
Corydalis davidii	Shrub	Ranunculus sikkimensis	Herb
Corylopsis himalayana	Tree	Rhododendron abroreum	Tree
Costus speciosus	Shrub	Rhododendron grande	Tree
Costus variegata	Shrub	Rhododendron chamaethomsonii	Srub
Cryptomaria japonica	Tree	Rhododendron falconeri	Tree
Cupresus tolulata	Tree	Rhododendron d	Shrub
Cymbidium sp.	Herb	Rhus simialata	Tree
Chimonobambusa collosa	Tree	Rhus griffithi	Tree
Daphnae cannabina	Shrub	Rubia cordiafolia	Herb
Dendrobium moschatum	Herb	Rubus ellipticus	Shrub
Dendrocalamus hamiltonii	Herb	Rubus hypragyrus	Tree
Drymaria cordata	Shrub	Rubus nepalensis	Tree
Duabanga grandiflora	Tree	Salix sp.	Tree
Elaeocarpous ganitrus	Tree	Sapium insignae	Tree
Eleagnus conferta	Shrub	Schima khasiana	Tree
Entada purseatha	Herb	Schima wallichii	Tree
Eriobotrya benghalensis	Tree	Sellaginella sp.	Herb
Erythrina indica	Tree	Senecio wallihi	Herb
Eryugium foetidium	Herb	Smilax megacarpa	Herb
Eugenia jambolana	Tree	Spondis auxalaris	Tree
Eurya acuminate	Tree	Sterculia villosa	Shrub
Ficus reliogiosa	Tree	Swertia chyrita	Herb
Ficus glomerata	Tree	Symplocos alternata	Tree
Gambleia ciliate	Tree	Symplocos spicata	Tree
Gaultheria fragratissima	Tree	Syzygium formosum	Tree
Hypericum uralum	Shrub	Taxus wallichiana	Tree
Hypericum uralum	Herb	Thalictrum foliolosum	Herb
Illicium griffith	Tree	Thuja orientalis	Tree
Juglans regia	Tree	Thuja roxburghii	Tree
Juniperous squamata	Tree	Thysaloma maxima	Shrub
Knema latifolia	Tree	Tsuga dumosa	Tree
Lagerostromia flosreginae	Tree	Vibranum sp.	Tree
Lindera pulcherima	Tree	Viola sikkimensis	Herb
Litsea elogata	Tree	Woodfordia sp.	Shrub
Lyonia ovalifolia	Tree	Zanthoxylum oxyphylum	Tree
Magnolia griffithi	Tree	Ziziphus jujubae	Tree
Mahonia nepalensis	Shrub	Ziziphus apetala	Shrub
Mangifera sylvatica	Tree		

There is now growing consensus among conservation planners that forest patches such as sacred groves are likely to be the key to maintaining biodiversity in the increasingly urbanized world. In recent years, the conservation community has realized that the long-term survival of biodiversity depends on the effectiveness with which such forest remnants can be managed (Bhagwat et al., 2005). While Colding and Folke (2001) proposed that conservation planners should devote careful consideration to already existing, local, informal institutions and involve local people in planning. Berkes (2004) also argued that there has been a shift in ecology and applied ecology towards a systems view on the environment, a perspective that sees humans as a part of the ecosystem.

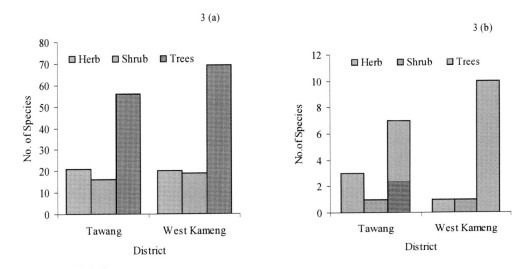

Figure 3. Distribution of (a) total plant species and (b) rare plant species in the sacred groves of Tawang and West Kameng districts of Arunachal Pradesh.

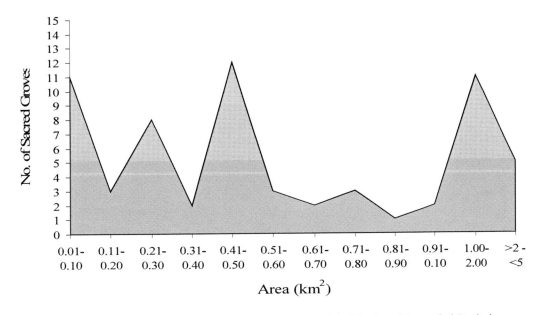

Figure 4. Area-wise distributions of sacred groves in the high altitudes of Arunachal Pradesh.

Tourism Potential of These Sacred Groves

These informal aspects of community based conservation have kept the large area under conservation network outside the national conservation planning framework. Unlike the other protected areas in the region, many of the biologically richest ecosystems such as sacred groves are poorly suited to ecotourism development because of factors such as difficult access, intangible wildlife, lack of information, etc. However, they can also generate good revenue to provide an effective incentive for conservation in areas where there is strong pressure on land and biological resources. There is no doubt that many community-based ecotourism projects create some local employment or generate some revenues that enhance some local incomes or help support community projects (Kiss, 2004). In many countries community-based ecotourism has become a popular tool for biodiversity conservation; based on the principle that biodiversity must pay for itself by generating economic benefits, particularly for local people (Kiss, 2004). Ecotourism development has become a prominent approach to address socioeconomic concerns along with the conservation and sustainable development. Hence, in this region ecotourism can also contribute both to conservation and rural development by generating revenue for these sacred groves management and by providing local communities with sustainable livelihood alternatives and economic benefits.

Moreover, little tourism in the region has been under outsider control and ownership, and most of the revenue is generated outside the local economy, while revenue generated at the local economic level does not benefit communities living in the area. To change this scenario would require focusing on developing the capacity of residents to become involved in tourism planning, management and operation. More significantly, as the poor and disadvantaged community members suffer most under resource use restrictions, they should become the beneficiaries of ecotourism development. With the involvement of local people in tourism development and the provision of economic benefits from tourism would thus be crucial steps to meet the subsistence and livelihood needs of communities living in the area. All community-based ecotourism initiatives should be centered on a clear strategy settled and understood by the local community and all other stakeholders with a consciousness in tourism and conservation. The strategy should enable an inclusive representation to be formed of needs and opportunities in an area, so that a range of balancing actions can be taken. Appropriate recreational and special awareness activities, such as trail walking, trekking, photography and participatory conservation programmes, may also be part of ecotourism. In some locations, hunting and fishing may be included as appropriate activities, provided that they are carefully explored and controlled within a management plan that supports conservation. This kind of sustainable use relies on local knowledge, provides significant local income and encourages communities to place a high value on wildlife, resulting in outstanding conservation benefits (Ashley and Garland, 1994; Wells, 1997; Wunder, 2000). One of the main benefits from working on a strategy is to provide the community with the tools and knowledge necessary for decision making. The approach should be community-based and community-focused. However, it is indispensable and as far as possible local community with experience and knowledge of tourism and conservation should involve in its execution. People involved should include representatives of the local community, knowledgeable tourism operators, local entrepreneurs, relevant NGOs, conservation agencies like WWF, WII, etc., including protected area managers, and local government authorities. Links should be made as appropriate to the local, regional, national and international levels.

On the other hand, the cultural heritage and ethnicity of the region should not be disturbed and, if possible, should be enhanced by tourism. Ecotourism should encourage people to value their own cultural heritage. However, culture is not static and communities may wish to see change. In order to minimize economic leakage, every effort should be made to use local produce and services, and to favour the employment of local people. This may require action to identify local, sustainable sources. Producers can be assisted through the formation of local groups and networks and help with contacts, marketing and pricing. Local communities should be encouraged and helped to take account of these issues themselves without any effect on their living standards, through information, training and demonstration.

Conclusion

The processes of modernization, the advent of modern religious faith and the spread of education have changed the lifestyle of traditional societies and thus have considerably eroded the traditional beliefs of the indigenous resident communities regarding the sacredness of these relict patches. Nowadays rituals, beliefs and other religious ceremonies associated with these sacred forests are looked after by the poor and illiterate and aged populations of the society. Apart from these, the vegetation structure of the many sacred groves has also been changing through natural calamities like landslides, heavy snowfalls, storms and other human-oriented anthropogenic disturbances like home building, road construction, timber and fuel wood collection and other developmental activities. This has led to a reduction in the size and number of sacred groves. On the other hand, in many villages sacred groves still exist, but the protection and management, worship and associated rituals are no longer performed due to the effects of modernization; thus, a rich cultural heritage is eroding day by day, and these sacred groves with their dense primary vegetation and high species richness stand merely as a symbol of their old sacredness and beliefs. The degradation of sacred groves not only signifies the loss of species-rich relict vegetation and animals, but also the loss of rich cultural heritage of the region. Considering the various dimensions of the sacred groves, it is clear that these relict forest patches need proper conservation and protection by formulating consistent conservation strategies in order to save them from further degradation, at least in the mega diversity hotspot region. One of the effective strategies could be provision of incentives to those societies involved in conservation, and by effective human-centered management strategies involving all cross-sections of human society.

Acknowledgements

This research is supported by the Department of Scientific and Industrial Research, Government of India, New Delhi. Thanks are given to the villagers of West Kameng and Tawang districts of Arunachal Pradesh for their cooperation and support.

References

Anonymous. (2004). *Statistical abstract of Arunachal Pradesh*. Directorate of Economics and Statistics. Government of Arunachal Pradesh, Itanagar, India.
Ashley, C. & Garland, E. (1994). Promoting community-based tourism development: what, why and how? Research Discussion Paper No.4. Department of Environmental Affairs, *Ministry of Environment and Tourism*, Namibia.

Berkes, F. (2004). Rethinking community-based conservation. *Conservation Biology* **18**(3): 621-630.

Bhagwat. S. A., Kushalappa. C. G., Williams, P. H. & Brown, N. D. (2005). The role of informal protected areas in maintaining biodiversity in the Western Ghats of India. *Ecology and Society* **10** (1): 8.

Chhetri, D. R., Basnet, D., Chiu, P. F., Kalikotay, S., Chhetri, G. & Parajuli, S. (2005). Current status of ethnomedicinal plants in the Darjeeling Himalaya. *Current Science* **89** (2): 264-268.

Colding, J. & Folke, C. (2001). Social taboos, invisible system of local resource management and biological conservation. *Ecological Applications* **11**(2): 584-600.

Gadgil, M. & Vartak, V. D. (1975). Sacred Groves of India: A Plea for continued conservation. *Journal of Bombay Natural History Society* **72**: 314-320.

Khan, M. L., Menon, S., Bawa, K. S. (1997). Effectiveness of the protected area network in biodiversity conservation: A case-study of Meghalaya state. *Biodiversity and Conservation* **6**: 853-868.

Kiss, A. (2004). Is community-based ecotourism a good use of biodiversity conservation funds? *Trends in Ecology and Evolution* **19** (5): 232-237.

Khumbongmayum, A. D., Khan, M. L., Tripathi, R. S. (2005). Sacred groves of Manipur Northeast India: biodiversity value, structure and strategies for their conservation. *Biodiversity and Conservation* **14**: 1541-1582.

Khumbongmayum, A. D., Khan, M. L., Tripathi, R. S. (2004). Sacred groves of Manipur-ideal centers for biodiversity conservation. Current Science **87** (4): 430-433.

Kosambi, D. D. (1962). *Myth and Reality*. Popular Press, Bombay, India.

Lama, T. (1999). *The Monpas of Tawang: a profile*. Himalayan Publishers, Itanagar, Arunachal Pradesh, India.

Malhotra, K. C., Ghokhale, Y., Chatterjee, S., Srivastava, S. (2001). *Cultural and Ecological Dimensions of Sacred Groves in India*, Indian National Science Academy Publication, New Delhi.

Ramakrishnan, P. S. (1998). Conserving the Sacred: Where do we stand? In: Ramakrishnan, P. S, Saxena, K. G, Chandrashekhara, U. M. (Editors). *Conserving the Sacred for Biodiversity Management*. Oxford and IBH Publishing Co. Pvt. Ltd. pp. 439-455.

Roy Burman, J. J. (1992). The institution of sacred grove. *Journal of Indian Anthropological Society* **27**: 219-238.

Takhtajan, A. (1969). Flowering Plants: Origin and Dispersal. Oliver and Boyd, Edinburgh.

Tiwari, B. K, Barik, S. K, Tripathi, R. S. (1999). *Sacred Forests of Meghalaya. Biological and cultural diversity*. Regional Centre, National Afforestation and Eco-Development Board, North- Eastern Hill University, Shillong, India.

Wells, M. (1997). Economic Perspectives on Nature Tourism, Conservation and Development. *Environment Department Paper No. 55* (Environmental Economics Series), The World Bank.

Wunder, S. (2000). Ecotourism and economic incentives: an empirical approach. *Ecological Economics* **32**: 465–479.

In: National Parks: Biodiversity, Conservation and Tourism
Editors: A. O'Reilly and D. Murphy, pp. 147-159
ISBN: 978-1-60741-465-0
© 2010 Nova Science Publishers, Inc.

Chapter 8

HIGH TEMPERATURE ENVIRONMENTS REPRESENT A MODEL FOR THE ANALYSIS OF BACTERIAL DYNAMICS AND PRESERVATION OF NATIONAL PARKS

M.C. Portillo and J.M. Gonzalez
IRNAS-CSIC, Sevilla, Spain

Abstract

Several National Parks located in different countries hold high temperature environments such as hot springs or volcanic sites. Among them, Teide National Park and Timanfaya National Park in Canary Islands (Spain), Kamchatka National Park in the Peninsula of Kamchatka (Russia), or Yellowstone National Park (USA) are some well known parks with hot environments. Microbial community studies at these sites have reported the presence of high and low temperature microorganisms which could show critical information on the exposure of these parks to diverse allocthonous microorganisms. Since high temperature environments are prohibitive to mesophilic, temperate microorganisms, the detection of these microorganisms indicates elevated dispersion rates and suggests an incredible colonizing potential. Similar results have been obtained in several National Parks from different world locations corroborating these findings. Since microorganisms play essential roles in biogeochemical cycles of major elements, maintaining their functionality and diversity represent decisive aspects for preserving the equilibrium of their communities and consequently conserve the integrity and variety of environments and landscapes protected within the National Parks.

Introduction

One of the major problems concerning National Parks is how to preserve their landscapes and biodiversity. While the diversity of plants and animals can be experimentally assessed, their protection involves the maintenance of their ecosystems and periodic monitoring. Any change in an environment can certainly have some effect on the plants and animals living there and so, the consequences of changes at a variety of scales is hard to predict although variations often lead to a reduction or homogenization of animal and plant diversity (Redford and Brosius, 2006). A search for models of study has been a target of scientists involved in that type of analyses. Nevertheless, while macroorganisms are often studied in National Parks and their preservation seriously considered, microorganisms are frequently overlooked and forgotten despite their critical contribution to ecosystem maintenance.

Microorganisms are those living beings too small to be seen by the naked-eye. The need for a microscope is required for their direct observation. Microorganisms comprise drastically different types of organisms such as some Eukaryotes (including fungi, protozoa, among other groups), Bacteria and Archaea. Bacteria and Archaea constitute the group named Prokaryotes which is, simplistically, defined as those cells lacking cellular nucleus. All the others, the Eukaryotes, are formed by cells relatively similar to those of animal and plant tissues and have nucleus. While Bacteria and Archaea might look like belonging to the same group, their physiological and genetic properties are radically different (Madigan et al. 2008). For a long time, Archaea have been considered as characteristics of extreme environments such as high temperature, high salinity, extremely acid environments. Today, we know the Archaea are present in any environment although it is generally believed that Bacteria are dominant in most temperate systems (Aller and Kemp, 2008).

National Parks represent places of great interest requiring to be preserved for future generations. National Parks located around the World represent a large variety of environments, and hold unique animals and plants, some of them close to the extinction. These sites require strict conservation strategies and a total isolation from human intervention. Among these parks a few of them present a number of high temperature environments. Some typical National Parks well known by their high temperature sites and with obvious volcanic origin are, for example, Yellowstone National Parks (USA), Kamchatka National Park (Russia), Teide National Park and Timanfaya National Park (Spain). These environments could be represented by hot springs or high temperature anomalies, generally related to volcanic environments. Temperatures near to 100°C can be detected. At these temperatures, closed to boiling water, only prokaryotic cells are able to develop. A gradient of temperatures creates adequate environments for many other microorganisms representing sites showing a large microbial diversity. Of course, most microorganisms grow under temperate conditions but some have been described to require high temperature environments (Stetter, 1999). High temperature requiring microorganisms are named thermophiles, extreme thermophiles or hyperthermophiles, depending on their preferred growth temperature, >50°C, >70°C, or >85°C, respectively (Blöchl, 1995; Madigan et al. 2008). These thermophiles are adapted to grow at high temperatures and are unable to develop at temperate environments. Mesophiles, on the contrary, are those microorganisms able to grow under temperate conditions and are unable to survive under high temperatures.

Since mesophiles, most known microorganisms, do not survive under high temperature conditions and thermophiles are the only ones adapted to high temperature, the microorganisms found at high temperature environments should belong to that niche or be cells just arrived to the hot spot from a different origin. In the last scenario, high temperature environments, could represent model systems for analyzing the arrival of microorganisms to specific sites and this case could be used for monitoring microbial immigration and dispersion, which are critical parameters to study bacterial population dynamics. This study focuses on experimental procedures to assess microbial communities considering high temperature environments as model study sites. The model environments can be used for analyzing the diversity present at these sites and their consequences for biodiversity preservation and National Park conservation.

Methods

Sampling Sites

Three National Parks were studied, Teide National Park (Tenerife Island, Canary Islands, Spain), Timanfaya National Park (Lanzarote Island, Canary Islands, Spain) and Kamchatka National Park (Peninsula de Kamchatka, Russia) for presenting high temperature environments. Information on these environments has been provided in previous publications (see below) and their location in a World map is presented in Figure 1. Briefly, these environments are characterized as described below.

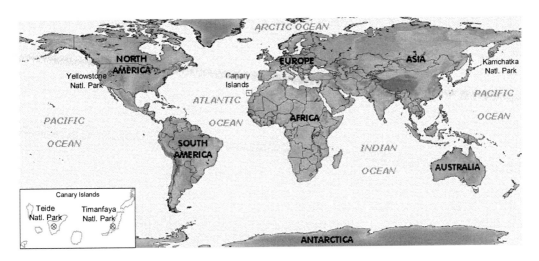

Figure 1. World map showing the location of the National Parks mentioned in this study, Teide National Park and Timanfaya National Park in Canary Islands (Spain), Kamchatka National park (Russia), and Yellowstone National Park (USA).

The archipelago of Canary Islands (Spain) is located just over 110 Km off the Northwestern African coast and the Sahara Desert. A description of the volcanic origin of these islands has been previously reported (Rothe, 1974; Mezcua et al. 1992; Carracedo, 1999). Teide National Park is located in Tenerife Island, one of the largest islands forming the Canary Islands. Teide is the major volvano at this island which remains active. The highest summit of the park is the Teide Volcano. Near the summit, it shows some spots at high-temperature (around 70°C). Timanfaya National Park presents high temperature spots, named thermal anomalies, which have low water content, and temperatures between 70 to 95°C. A few meters below ground it is possible to detect temperatures up to 400°C. Solidified lava rivers can be observed at these two parks. Generally, these parks present dry environments with high wind rate, mostly coming from Africa, especially from the Sahara Desert. No visible biological colonization (i.e., plants or liquens) was observed at the sampled sites which were composed mostly of basaltic and pyroclastic rocks of different sizes. These areas showed a low water content, with desert-like xeric environments of volcanic rocks.

Kamchatka National Park is located at the Pacific Ocean coast of Russia, and presents numerous hot springs, at temperatures up to around 90°C, and other volcanic environments. Previous descriptions of this park have been reported (Bindeman et al. 2004). Unlike the parks in Canary Islands, a number of thermophiles have been isolated from Kamchatka (Sokolova et al. 2004; Prokofeva et al. 2005; Kevbrin et al. 2005) while, to our knowledge, no high-temperature isolates are available from Canary Islands.

Sampling Conditions and Sample Preservation

Samples for the detection of microorganisms based on their nucleic acid information were collected from the surface into 1.5-ml microtubes, preserved in solution RNA-latter and immediately closed. Samples were maintained at the lowest temperature possible (on ice during transportation to the lab and at -80°C afterwards). Samples for culturing microorganisms were preserved at 4°C until arrival to the laboratory when they were immediately processed. Temperature was measured at surface using a K-thermocouple thermometer. Sampled sites showed stable high temperature throughout the year (Araña et al. 1984).

Nucleic Acids Extraction and Processing

Based on the DNA and RNA, microorganisms can be classified in their taxonomical groups. Using the 16S ribosomal RNA gene as a target site for experimental bacterial classification, it is possible to detect microorganisms without a need for culturing them. This is a great advantage since most microorganisms in natural environments are non-culturable, and the total cultured microorganisms represent well less than 1% of total microbial abundance (Ward et al. 1990). Thus, the molecular detection and classification of bacteria, represent a culture-independent approach to avoid culturing biases. Since the RNA content in microbial cells is proportional to the metabolic activity of a cell, the detection of bacteria in base to their RNA implies that the microorganisms found are among those developing highest metabolic activity. Unlike RNA, DNA is independent of metabolic activity, and the bacteria detected by DNA analysis will represent those microorganisms present in the studied environment independently if they develop or carry out any metabolic activity *in situ*. Thus, using DNA or RNA, we can assess the metabolically active bacteria in an environment (based on RNA) and those bacteria present in the ecosystem independently of their *in situ* metabolism.

DNA was extracted using the Nucleospin Food DNA extraction kit (Mackerey-Nagel, Düren, Germany) following the manufacturer's recommendation. Total RNA was extracted with the RNAqueous4PCR kit (Ambion, Austin, USA). This last protocol included a DNaseI treatment (37°C for 1 hour) to remove DNA traces copurified during the extraction procedure. Numerous controls without sample (as controls from reagents and tubes) were carried out in parallel to the samples.

DNA Amplification and Reverse Transcription

Complementary DNA to the 16S rRNA genes to be amplified were obtained by reverse transcription using Thermoscript (Invotrogen, Carlsbag, CA, USA). This reaction was performed at 55°C for 1 h with the 16S rRNA gene-specific primer 518R (5'-ATT ACC GCG GCT GCT GG; Gonzalez et al. 2005a). A variety of controls were carried out including those lacking reverse transcriptase, and lacking RNA.

PCR amplification was performed using the primer pairs 341F (5'-CCT ACG GGA GGC AGC AG; Gonzalez et al. 2005a) and 518R from reverse transcribed DNA, and 27F (5'-AGA GTT TGA TYM TGG CTC AG) and 907R (5'-CCC CGT CAA TTC ATT TGA GTT T) from extracted DNA, and the following thermal conditions: 95°C for 2 min; 30 cycles of 95°C for 15 s, 55°C for 15 s, and 72°C for 1 min; and a final incubation at 72°C for 10 min.

Environmental Clone Libraries

Products of amplification were purified using a JetQuick Kit (Germany). And cloned with the TOPO-TA cloning kit (Invitrogen). These 16S rRNA libraries were screened to select unique clones as previously described (Gonzalez et al. 2003). Selected clones were sequenced to identify some of the microorganisms present in the studied samples.

Microbial Community Fingerprints

In order to obtain a visual representation of the major components of the microorganisms present in the studied samples, fingerprinting analysis was performed by Denaturing Gradient Gel Electrophoresis (DGGE) (Gonzalez and Saiz-Jimenez, 2004). For this analysis, the primers 341F-GC (supplemented with a GC-rich tail) and 518R were used. Differences between molecular fingerprints were estimated following Portillo and Gonzalez (2008).

Sequence Analysis

Search for the closest microorganisms from the DNA database to the detected sequences was performed using the blastn algorithm (http://www.ncbi.nlm.nih.gov/Blast/; Altschul et al. 1990). Sequences were analyzed for the presence of chimeras following the procedure of Gonzalez et al. (2005b)

RNA Decay Analysis

In order to determine the time frame available for the detection of microorganisms after arrival to high temperature sites, mesophilic bacteria isolated from the studied environments and culture collection strains were assayed to measure the period of time available for the detection of those bacteria in base to the remaining RNA. The method was developed as a modification of the meting curves procedure described by Gonzalez and Saiz-Jimenez (2002). Briefly, the method stains RNA using the fluorescent dye RiboGreen (Molecular Probes,

Eugene, Oregon) which is specific for RNA staining. Decay of fluorescent signal overtime at 80°C was recorded using an iQ iCycler optical thermocycler (Bio-Rad, Hercules, California). T90 values for the tested bacteria were estimated as the time when 90% of the initial RNA was degraded.

Results

During this study, most detected bacteria belong to well described groups which have no thermophilic relatives. Thus, most bacteria detected in the studied high-temperature samples were mesophiles which was also confirmed since cultures obtained from the same sites only showed growth below 50°C. The only exception was a relative to *Streptococcus thermophilus* detected from DNA-based analyses in a sample from Timanfaya National Park. The mesophilic nature of the detected bacteria can be observed from the list of the taxonomic affiliation of detected bacteria in the three studied sites and their expected temperature range of growth presented in Table 1.

Table 1. List of the major bacteria detected in high temperature samples, their taxonomic affiliation, optimum temperature range for growth, the origin of the sample were it was found, and whether the molecular survey was carried out in base to DNA and/or RNA.

Taxonomic afiliation	Optimum temperature range	Sample origin	DNA or RNA
Acidobacteria			
Acidobacterium	15-22	Can	DNA
Bacteroidetes			
Cytophaga	25-28	Can	RNA
Cryomorphaceae	28-35	Can	RNA
Actinobacteria			
Actinomycetes	35-39	Can	DNA
Propionibacterium	25-32	Can	DNA, RNA
Corynebacterium	30-35	Can	RNA
Microbacterium	28-35	Can	RNA
Cellulomonadaceae	38-40	Can	DNA
Firmicutes			
Enterococcus	35-40	Can	DNA
Streptococcus thermophilus	40-45	Can	DNA
Bacillus	30-40	Can	RNA
Sporolactobacillus	35-39	Kam	DNA
Clostridium	30-40	Kam	DNA
Alpha-Proteobacteria			
Methylobacterium	28-35	Can	DNA
Bosea	20-28	Can	RNA
Sphingomonas	30-35	Can, Kam	DNA, RNA
Hyphomonas	22-37	Can	RNA
Caulobacter	25-30	Can	RNA
Mesorhizobium	25-30	Kam	RNA

Table 1. Continued.

Taxonomic afiliation	Optimum temperature range	Sample origin	DNA or RNA
Bradyrhizobium	25-30	Kam	RNA
Beta-Proteobacteria			
Variovorax	20-25	Can	RNA
Burkholderia	30-35	Can, Kam	DNA, RNA
Epsilon-Proteobacteria			
Campilobacter	30-45	Can	RNA
Gamma-Proteobacteria			
Pseudomonas	34-40	Can, Kam	DNA, RNA
Steroidobacter	25-30	Can	DNA
Xanthomonas	26-30	Can	RNA
Photorhabdus	25-30	Can	RNA
Shigella	35-40	Can	RNA
Acinetobacter	5-45	Kam	DNA
Verrucomicrobia			
Verrucomicrobia	26-33	Can	DNA

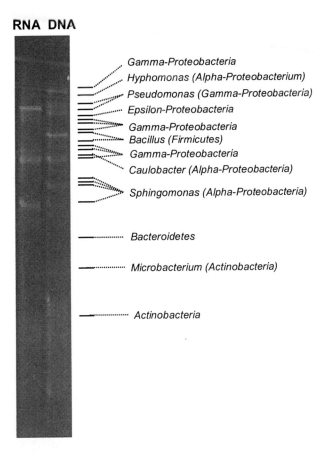

Figure 2. Bacterial community fingerprint performed by DGGE based on DNA and RNA from Teide National Park samples at 70°C. The identification of the microorganism corresponding to each band is indicated with the percentage of similarity to its closest homologue from GenBank based on the 16S rRNA gene sequence.

Molecular fingerprints of the bacterial communities detected in samples from the Teide and Timanfaya National Parks are shown in Figures 2 and 3, respectively, showing the location of the bands corresponding to each of the detected bacteria according to their migration during electrophoresis by DGGE. Significant differences (P<0.001) between DNA and RNA-based analyses were observed. Besides, differences were also observed (P<0.001) between the banding patterns detected for DNA in Teide and Timanfaya National Parks and between RNA analysis from these two parks. These differences of community fingerprints clearly determine a lack of contamination during laboratory handling since all these analyses were carried out simultaneously.

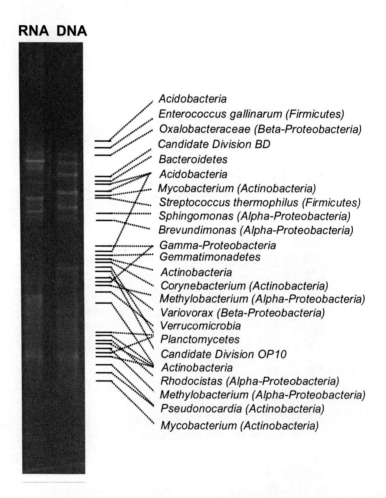

Figure 3. Bacterial community fingerprint performed by DGGE based on DNA and RNA from Timanfaya National Park samples at 90ºC. The identification of the microorganism corresponding to each band is indicated with the percentage of similarity to its closest homologue from GenBank based on the 16S rRNA gene sequence.

Comparing the major bacterial components detected in the samples from the national parks of Kamchatka and those of Canary Islands, we observed that a relatively similar distribution is observed (Figure 4). For instance, a similar percentage (23%) of total detected bacteria belong to the Alphaproteobacteria and Gammaproteobacteria in both geographically distant sites. Firmicutes and Betaproteobacteria were also well represented in high temperature samples from Kamchatka and Canary Islands. While many Firmicutes could actually form spores as survival mechanisms against adverse conditions, the Proteobacteria represent the majority of detected mesophiles and they are not able to form spores. Besides, the detection in base to RNA also indicate that these bacteria are living forms and not resting stages (i.e. spores).

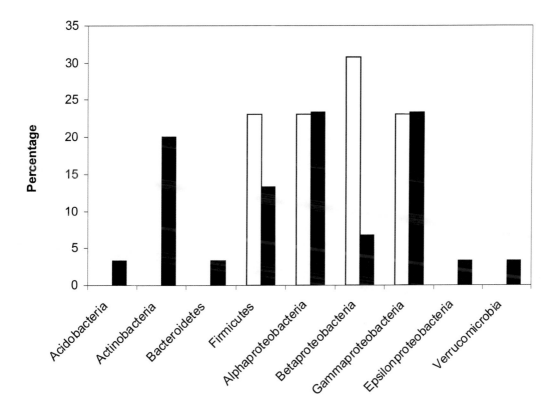

Figure 4. Proportion of mesophiles belonging to major bacterial groups detected in high temperature samples from the Canary Islands (black bars) and Peninsula of Kamchatka (white bars).

The decay rate of mesophilic bacteria RNA at high temperature (80°C) was expressed by the T90 of RNA from *Escherichia coli* (Gammaproteobacteria) and a *Sinobacter* sp. (Betaproteobacteria), an isolate from high temperature samples of Canary Islands. The decay data for RNA of these two bacteria are shown in Figure 5. T90 values obtained from these two cases were 6 min (for the *Sinobacter* strain) and 3 min (for *E. coli*). This implies that 90% of RNA is degraded to nucleotides in a time frame of a few minutes of exposure to high temperatures. Thus, the time available for the detection of these bacteria in base to their RNA, and at those high temperatures, is just a few minutes.

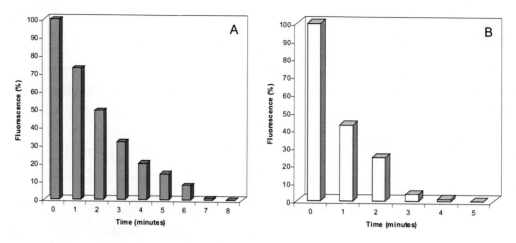

Figure 5. Results from RNA decay experiments carried out for *Escherichia coli* (A) and an isolate from high temperature environments at Canary Islands (B), *Sinobacter* sp. (Betaproteobacteria).

Discussion

Microbial dispersion is an interesting phenomenon still to be understood. While it is known that wind forces are important factors affecting the mobility of airborne microorganisms (Karra and Katsivela, 2007) as well as spores and pollen (Jones and Harrison, 2003), the level of dispersion and the significance of this process on the dynamics of microbial communities is not clear.

Most commonly found microorganisms present optimum temperature for growth under temperate conditions. Thermophiles are an exception, generally, requiring temperatures higher than 50°C. Mesophiles survive very short times to high temperature exposures. Thus, high temperature environment might represent natural, model sites for assessing the frequency and rates of arrival of microorganisms to new environments.

The results presented in this study confirm that mesophilic microorganisms are actually arriving to hot spots where they can not survive. Survival time is within a few minutes as well as the time available before these microorganisms are undetectable using RNA analyses. Both mesophilic microbial cells and their nucleic acids are quickly degraded under high temperature conditions. Only thermophiles could live and develop at high temperatures. Independently of the finding of thermophiles in the studied samples, a major concern is the continuous arrival of microorganisms to these high temperature environments.

Incredible high rates of arrival of microorganisms to specific systems are deduced from the results obtained in this study. A number of different microbial groups, including the most frequent bacterial Divisions such as Proteobacteria, Firmicutes, Actinobacteria, among others, were detected at every sample analyzed during this study. Consequently, immigration of microorganisms is a constant in the studied model systems and so, in most natural environments. Thus, the detected microbial communities contain a high proportion of allochtonous microorganisms. This brings up a question on the actual composition of the authochtonous microorganisms in a given environment. From these results, one could assume that the immigrant component might be a significant fraction of the total detectable microbes in a given environment.

While the rate of arrival of microbial immigrants is very high further experimental work is require to estimate accurate values and relate those data to environmental factors. However, our results indicate a constant arrival of microorganisms to an ecosystem. Thus, there is a high risk of novel microbial colonizations or incorporations to a system. A consequence of this risk is that novel processes or transformations of the standard behaviour of a microbial community can occur. Microbial immigration can generate changes in the communities of a given ecosystem. Since microorganisms are critical components of our ecosystems, changes in the behaviour of these communities might be reflected in deviations of the standard ecosystem processes.

Microbial diversity have been reported to be huge (Curtis et al 2002). The consequences of this diversity in ecosystem variability remain to be understood. The presented results suggest that microbial diversity at a local scale (i.e., within an ecosystem) might be much lower than previously proposed (Whitman et al. 1998) because those microorganisms just arrived to the system might not be adapted to survive in it and will be eliminated soon after. This is the case of hot environments, where most arriving microorganisms will not have any opportunity to survive and develop. However, these results also suggest that microorganisms can go almost everywhere. Thus, microorganisms can reach any place on Earth, and the environment will select among them. This is a variation of the classical hypothetical view of Baas-Becking (Baas-Becking, 1934) which states "microorganisms are everywhere, (but) the environment selects". Nevertheless, the presented results are in agreement with the current estimates of a huge microbial diversity in our planet.

Huge microbial diversity estimates present important consequences for biogeography determinations and microbial diversity predictions. The evidence that microbial diversity, if reduced through human intervention, could be naturally restated presents a positive perspective. Nevertheless, microbial diversity is too high to be experimentally determined and microbial biogeography is difficult to define. These two aspects are of great interest on the preservation of microbial communities which constitutes a critical component of the ecosystems to be preserved in National Parks.

Conclusion

High temperature model ecosystems have been used to approach the importance of microbial dispersion and immigration. The level of microbial diversity in our planet remains to be determined. The conservation of National Parks and their unique ecosystems and landscapes should avoid drastic changes of their microbial diversity and their role in these systems. Microbial community dynamism is a field that needs further investigation and their consequences might have direct effects on conservation strategies in order to maintain the uniqueness of National Parks and avoid processes of system homogenization in our planet.

Acknowledgements

This research has been supported by the Spanish Ministry of Education and Science (REN2002-00041) and the Andalusian Government (Bio288).

References

Aller, J. Y. & Kemp, P. F. (2008). Are Archaea inherently less diverse than Bacteria in the same environments? *FEMS Microbiology Ecology*, **65**, 74-87.

Altschul, S. F., Gish, W., Miller, W., Myers, E. W. & Lipman, D. J. (1990). Basic local alignment search tool. *Journal of Molecular Biology*, **215**, 403-410.

Araña, V., Diez, J. L., Ortiz, R. & Yuguero, J. (1984). Convection of geothermal fluids in the Timanfaya volcanic area (Lanzarote, Canary Islands). *Bulletin of Volcanology* **47**, 667-677.

Baas-Becking, L. G. M. (1934). *Geologie of Inleiding Tot de Milieukunde*. The Hague: Stockum & Zoon,.

Bindeman, I. N., Ponomareva, V. V., Bailey, J. C. & Valley, J. W. (2004). Volcanic arc of Kamchatka: a province with high-$\delta^{18}O$ magma sources and large-scale $^{18}O/^{16}O$ depletion of the upper crust. *Geochimica and Cosmochimica Acta* ,**68**, 841-865.

Blöchl, E., Burggraf, S., Fiala, G., Lauerer, G., Huber, G., Huber, R., Rachel, R., Segerer, A. & Stetter, K.O., Völkl. (1995). Isolation, taxonomy and phylogeny of hyperthermophilic microorganisms. *World Journal of Microbiology and Biotechnology*, **11**, 9-16.

Carracedo, J. C. (1999). Growth, structure, instability and collapse of Canarian volcanoes and comparisons with Hawaiian volcanoes. *Journal of Volcanology and Geothermal Research*, **94**, 1-19.

Curtis, T. P., Sloan, W. T. & Scannell, J.W. (2002). Estimating prokaryotic diversity and its limits. *Proceedings of the National Academy of Science USA*, **99**, 10494-10499.

Gonzalez, J. M. & Saiz-Jimenez, C. (2002). A fluorimetric method for the estimation of G+C mol% content in microorganisms by thermal denaturation temperature. *Environmental Microbiology*, **4**, 770-773.

Gonzalez, J. M., Ortiz-Martinez, A., Gonzalez-delValle, M. A., Laiz, L. & Saiz-Jimenez, C. (2003). An efficient strategy for screening large cloned libraries of amplified 16S rDNA sequences from complex environmental communities. *Journal of Microbiological Methods*, **55**, 459-463.

Gonzalez, J. M. & Saiz-Jimenez, C. (2004). Microbial diversity in biodeteriorated monuments as studied by denaturing gradient gel electrophoresis. *Journal of Separation Science*, **27**, 174-180.

Gonzalez, J. M., Portillo, M. C. & Saiz-Jimenez, C. (2005a). Multiple displacement amplification as a pre-PCR reaction to process difficult to amplify samples and low copy number sequences from natural environments. *Environmental Microbiology*, **7**, 1024-1028.

Gonzalez, J. M., Zimmermann, J. & Saiz-Jimenez, C. (2005b). Evaluating putative chimeric sequences from PCR amplified products and other cross-over events. *Bioinformatics*, **21**, 333-337.

Jones, A. M. & Harrison, R. M. (2004). The effects of meteorological factors on atmospheric bioaerosol concentrations – a review. *Science of the Total Environment*, **326**, 151-180.

Karra, S. & Katsivela, E. (2007). Microorganisms in bioaerosol emissions from wastewater treatment plants during summer at a Mediterranean site. *Water Research*, **41**, 1355-1365.

Kevbrin, V., Zengler, K., Lysenko, A.M. & Wiegel, J. (2005). *Anoxybacillus kamchatkensis* sp. Nov., a novel thermophilic facultative aerobic bacterium with a broad pH optimum from the Geyser valley, Kamchatka. *Extremophiles*, **9**, 391-398.

Madigan, M. T., Martinko, J. M., Dunlap, P.V. & Clark, D. P. (2008). *Brock Biology of Microorganisms* (12th Edn). San Francisco, CA, USA: Benjamin Cummings Publishing Co.

Mezcua, J., Buforn, E., Udias, A. & Rueda, J. (1992). Seismotectonics of the Canary Islands. *Tectonophysics*, **208**, 447-452.

Portillo, M. C. & Gonzalez, J. M. (2008). Statistical differences between molecular fingerprints from microbial communities. *Antonie van Leeuwenhoek*, **94**, 157-163.

Prokofeva, M. I., Kublanov, I. V., Nercessian, O., Tourova, T. P., Kolganova, T. V., Lebedinsky, A. V., Bonch-Osmolovskaya, E. A., Spring, S. & Jeanthon, C. (2005). Cultivated anaerobic acidophilic/acidotolerant thermophiles from terrestrial and deep-sea hydrothermal habitats. *Extremophiles*, **9**, 437-448.

Redford, K. H. & Brosius, J. P. (2006). Diversity and homogenization in the endgame. *Global Environmental Change*, **16**, 317-319.

Rothe, P. (1974). Canary Islands – Origin and evolution. *Naturwissenschaften*, **61**, 526-533.

Sokolova, T. G., Gonzalez, J. M., Kostrikina, N. A., Chernyh, N. K., Slepova, T. V., Bonch-Osmolovskaya, E. A. & Robb, F. T. (2004). *International Journal of Systematic and Evolutionary Microbiology*, **54**, 2353-2359.

Stetter, K. O. (1999). Extremophiles and their adaptation to hot environments. *FEBS Letters*, **452**, 22-25.

Ward, D. M., Weller, R. & Bateson, M. M. (1990). 16S rRNA sequences reveal numerous uncultured microorganisms in a natural community. *Nature*, **345**, 63-65.

Whitman, W. B., Coleman, D. C. & Wiebe, W.J. (1998). Prokaryotes: The unseen majority. *Proceedings of the National Academy of Science USA*, **95**, 6578-6583.

//
Microbial Diversity Supporting Unique Ecosystems within National Parks. The Doñana National Park as an Example

M.C. Portillo[1], M. Reina[2], L. Serrano[2] and J.M. Gonzalez[1]

[1] IRNAS-CSIC, Sevilla, Spain
[2] Department of Plant Biology and Ecology, University of Sevilla, Sevilla, Spain

Abstract

National Parks represent unique sites in need of preservation for future generations. Although frequently overlooked, the role of microorganisms in the functioning and biogeochemical cycling of elements is essential to support the maintenance of these ecosystems, and consequently their preservations within National Parks. Recent studies have provided serious grounds to confirm that microbial communities present in natural environments are much more diverse than previously imagined. In this study, we present a case study of a singular environment, the freshwater ponds of the Doñana National Park (Spain). Differences between microbial communities developing in close proximity can be detected suggesting the existence of drastically distinctive niches and microhabitats at a reduced spatial scale. The role of the invisible microorganisms must be considered when managing park conservation and analyzing ecosystem long-term stability. Consequences of the interactive role of microorganisms and environmental and geochemical factors can lead to ecosystem changes with important consequences for nutrient cycling and availability. Preservation of the huge diversity of microbial life in National Parks is a must if the equilibrium and stability of these environments and ecosystems is to be preserved.

Introduction

Microorganisms represent a critical portion of natural ecosystems. Numerous processes are carried out exclusively by microorganisms and their role in nutrient cycling and biogeochemical transformations is essential for the functioning of nature (Whitman et al.,

1998; Nee, 2004). Despite this fact, at present, microbial distribution, metabolism, and interactions with the environment and with larger living beings are mostly unknown. While animals and plants are being censored for centuries, the variety and distribution of microorganisms have received scarce attention through history (Nee, 2004).

One of the factors inhibiting the progress of environmental microbiology is the methodology. Microorganisms are too small to be studied with the naked eye; rather the use of microscopes is required to visualize single microbial cells. Besides microscopy, classic microbiological methods to make microbes visible required their culture on liquid or solid medium and so one could see either turbidity or the formation of colonies as a result of microbial growth. In order to observe microbial growth, an unbelievable high number of cells (reaching levels of over 10^{10} cells ml^{-1}) is needed in the cultures. In this way, culturing cells have been the method used to enumerate microorganisms from the environment (Madigan et al., 2008). Culturing allows knowing, for instance, under which conditions the cultivated microorganisms grow, which substrates can be used and which products are released under laboratory conditions. Nevertheless, by the last quarter of the previous century, scientists realized that culturing microorganisms introduced a huge bias in the counts of microorganisms. Only those microorganisms finding adequate conditions in the laboratory media could develop and form colonies. In fact, at present, microbiologists accept that only less than 1% of the microorganisms present in any given environment can be cultured (Ward et al., 1990). Consequently, most microorganisms would remain undetected if only culturing methods would be used.

The introduction of molecular methods, based on nucleic acid analysis, for the detection of microorganisms provided a huge advance in environmental microbiology (Pace, 1997). Today, we know than over a hundred bacterial divisions have been proposed and the diversity detected in our planet is unimaginable. The accepted fact is that microorganisms are too diverse to be experimentally evaluated (Curtis et al., 2002). Providing a number of the different microorganisms existing in our planet will only be based on speculation due to the incredible high numbers involved. Due to the problem to determine how many different microorganisms exist, the difficulty in their detection and classification, together to the complexity of structuring microbial communities in the environment, there is a great gap of information on the distribution and biogeography of microorganisms (Staley, 1997; Curtis and Sloan, 2005).

As a consequence of the latest developments in environmental microbiology, a long list of novel bacterial groups is being discovered. Woese (1987) proposed a description of bacteria divided in twelve divisions. Sixteen years later, Rappé and Giovannoni (2003) present an updated phylogenetic analysis of the domain Bacteria with a total of 52 divisions. Interestingly, only about half of them had, at least, one cultured representative. Today, over a hundred bacterial divisions have been proposed but most of them lack cultured representatives. Thus, most of bacterial divisions existing in our planet have never been cultured and thus, their metabolism remains unknown. This observation confirms the high bacterial diversity existing on Earth as well as the need for further investigation on microbial metabolism.

Despite the lack of information on the capabilities of a large fraction of the microorganisms present in nature, current knowledge allows confirming that the role of microorganisms in natural environments is critical to maintain the biogeochemical processes and the development of higher trophic levels. Thus, animals and plants can continue to develop and their ecosystems are maintained.

Microorganisms are essential for the maintenance of natural ecosystems. One of the routes to preserve unique environments, such as National Parks, would be to preserve the smallest creatures working on them, the microorganisms. Microbial diversity is huge, and there are reasonable pieces of evidence supporting that this diversity is a requirement for keeping a proper functioning of the environments despite any external influence. A large diversity, and a consequent rich genetic pool (Rauch and Bar-Yam, 2004), is required for microorganisms to face radical environmental changes and return the ecosystems to their previous state, re-establishing a previous equilibrium and contributing to maintain the environments for future generations to come. This study presents an example of the role of microorganisms and their diversity in a national park. The influence of microorganisms on the environment is analyzed.

Methods

Sampling Sites

The Doñana region extends along the coastal plain of the Gulf of Cadiz from the left bank of the Guadalquivir River Estuary to the Tinto River Estuary, and inland from the lower Guadalquivir River valley to the northern uplands bordering the Iberian Pyrite Belt (Figure 1). Doñana includes several territories with a different degree of environmental protection covering over 100 000 ha: a Biological Reserve created in 1964, a National Park that exhibits the highest degree of environmental protection in Spain (designated as a Biosphere Reserve in 1980, a Ramsar site in 1982 and a Natural World Heritage Site in 1995), and a Natural Park created as a surrounding protective area in 1989. This region has a Mediterranean climate with Atlantic influence, generally classified as dry subhumid. Rainfall is quite variable, both within a year and over the years, with a 580 mm yearly average, about 80% of which is distributed throughout a wet period from the end of September to the beginning of April. Summers are very dry and hot, while winters are short and mild. Water balance is generally deficient as rainfall exceeds evapotranspiration only during 3-4 months a year (Siljeström and Clemente, 1990). Three different landscapes are usually described: the marshland, the mobile dunes, and the stable sands (Figure 1). Hundreds of small ponds appear amid depressions of the stable sands when the water table rises above the topographical surface during heavy rains. The studied ponds range widely in size and depth according to seasonal variations: from over 10 ha and nearly 2 m deep during floods to complete desiccation after dry summers (Serrano et al., 2006). Samples were collected at different times of the year from the ponds named "La Dulce" and "Santa Olalla". At "La Dulce" Pond samples were collected at the West and East sites to compare the different microbial communities developing at these locations due to the clear visible differences observed between the environments at these two shores.

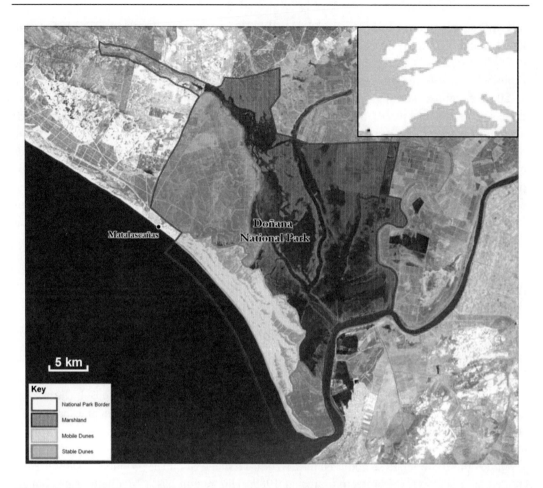

Figure 1. The Doñana National Park is located in Southern Spain, from the left bank of the Guadalquivir River.

Sampling Conditions and Sample Preservation

Samples for the detection of microorganisms based on their nucleic acid information were collected from the surface of sediments into 1.5-ml microtubes, preserved in solution RNA-latter (Ambion, CA, USA) and immediately closed. Samples were maintained at the lowest temperature possible (on ice during transportation to the lab and at -80°C afterwards) until processing.

Nucleic Acids Extraction and Processing

DNA was extracted using the Nucleospin Food DNA extraction kit (Mackerey-Nagel, Düren, Germany) following the manufacturer´s recommendations.

PCR amplification of 16S rRNA genes was performed using the primer pairs 27F (5'-AGA GTT TGA TYM TGG CTC AG; Gonzalez et al., 2005b) and 907R (5'-CCC CGT CAA TTC ATT TGA GTT T; Gonzalez et al. 2005b), and the following thermal conditions: 95°C for 2 min; 30 cycles of 95°C for 15 s, 55°C for 15 s, and 72°C for 1 min; and a final incubation at 72°C for 10 min. Products of amplification were purified using a JetQuick Kit (Germany), and cloned with the TOPO-TA cloning kit (Invitrogen). These 16S rRNA libraries were screened by DGGE analysis to select unique clones as previously described (Gonzalez et al., 2003). DGGE analyses are described below. Selected clones were sequenced by standard procedures to identify some of the microorganisms present in the studied samples.

In order to obtain a visual representation of the major components of the microorganisms present in the studied samples, fingerprinting analysis was performed by Denaturing Gradient Gel Electrophoresis (DGGE) (Gonzalez and Saiz-Jimenez, 2004). For this analysis, the primer pair 341F-GC (5'-CCT ACG GGA GGC AGC AG supplemented with a GC-rich tail) and 518R (5'-ATT ACC GCG GCT GCT GG; Gonzalez et al., 2005a) was used. Differences between molecular fingerprints were estimated using the software fingshuf following Portillo and Gonzalez (2008).

DNA sequences were individually inspected using the software Chromas version 1.45 (Technelysium, Tewantin, Australia). Edited sequences were used for a search of the closest homologue microorganisms to the ones corresponding to the sequences retrieved from the environment. This search was performed using the blastn algorithm (http://www.ncbi.nlm.nih.gov/Blast/; Altschul et al., 1990) and the GenBank DNA Database. Sequences were analyzed for the presence of chimeras following the procedure of Gonzalez et al. (2005b)

Results

Clear differences are obvious when comparing two sites of a relatively small freshwater pond at Doñana National Park (Figure 2). "La Dulce" Pond shows a West site of red-brownish color due to an iron-rich film covering the surface of water and shore sediment. This iron film is generated by the activity of microorganisms and directly affects the development of higher organisms (e.g., plants) on these areas when both shores are compared (Portillo et al., 2008).

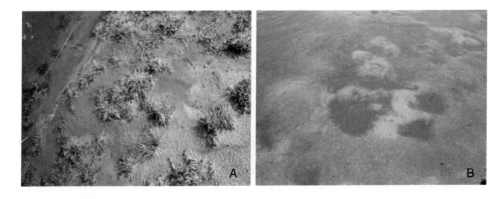

Figure 2. Different appearance of the two sites of "La Dulce" Pond at Doñana National Park. The West site (A) is covered by and Fe-rich film and the East site (B) showing a completely different environment and lacking that film.

A large variety of microorganisms have been detected in sediment samples from freshwater ponds at the Doñana National Park. A quick method to visualize the presence of a large number of microorganisms is provided by DGGE analyses, which allows characterizing a sample with its unique fingerprint (Figure 3). Figure 3 visualizes the high complexity of the studied communities at two different freshwater ponds within Doñana National Park. A comparison of microbial community fingerprints about a meter apart revealed no significant differences among their profiles. Comparison of microbial community fingerprints from two freshwater ponds in proximity (within a couple of kilometres) revealed significantly different profiles (P<0.001). This result indicates that microbial communities adapt to minor changes in the environment. Such environmental differences have been previously described (Serrano et al., 2006).

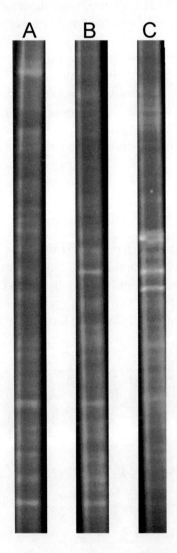

Figure 3. Microbial community fingerprints from samples of two freshwater ponds at Doñana National Park. "La Dulce" West site (A) covered by a Fe-rich film, "La Dulce" East site, and "Santa Olalla" (B) are the location of the compared samples.

A clear example of environmental differences is observed when comparing the two sites of a single freshwater pond as shown in Figure 2. The bacterial communities at these two shores are represented by their community fingerprints in Figure 3 (A and B). Thus, differences in these communities (P<0.001) directly lead to environmental changes. These changes are visible and generate two distinctive ecosystems. For instance, the iron-covered shore presents oxygen-depletion at the water column under the surface film while the eastern shore (lacking the iron film) shows this depletion only under the surface of the sediment layer. The microorganisms responsible for the formation of this iron film and the transformations required for that process to occur are strains of the Gammaproteobacteria genus *Enterobacter*, and the genera within the Firmicutes *Anaeroarcus* and *Clostridium*, as previously reported (Portillo et al., 2008).

Analyses of the components of a microbial community involves the identification of microorganisms by DNA-based molecular methods which is the most common culture-independent technique used for microbial surveys in the environment. Table 1 shows a list of some of the microbial groups more characteristics in the Doñana National Park since they are frequently detected in these environments. As shown, the metabolisms carried out by these different groups include a broad range of capabilities and, interestingly, most of these processes are unique to the microbial world, since animals and plants are unable to perform that type of transformations. For example, multiple processes are performed under aerobic and others under anaerobic conditions. While there are microorganisms that grow heterotrophically using a metabolism relatively similar to that of macroorganisms (i.e., animals), there are other microbes that obtain energy and carbon for growth from inorganic compounds. Sulfate-reducing and sulphur oxidizing bacteria carry out two critical processes within the biogeochemical cycle of sulphur in these environments. Nitrogen needs to be incorporated into biomass and nitrogen fixation from the atmosphere is carried out by some microorganisms. Also, microorganisms can transform nitrogen to any one of its natural forms, producing and consuming nitrate, nitrite, and ammonium and returning it to the atmosphere as N_2. Thus, microorganisms can carry out the whole cycle of nitrogen which is important to provide nutrients to the macroorganisms (i.e, plants). A specialized group of Archaea carries out a singular metabolism utilizing either carbon dioxide or acetate (or some other substrates) in the presence of hydrogen and producing methane. They are the methanogens which represent the last element of the organic carbon mineralization in the carbon cycle (Madigan et al., 2008). A specialized group is also constituted by the methanotrophic bacteria which oxidize methane producing organic matter. Another typical example of the role of microorganisms in nature is, for example, the cycle of iron. Microorganisms can reduce and oxidize iron, and use it when chelated with organic matter, closing the iron cycle within the microbial world. This is an example of the phenomenon observed in "La Dulce" Pond in Doñana National Park resulting in two completely different environments when the West and East shores are compared. The West site of the pond is generally covered by an iron film while the East site lacks it. Besides, the biogeochemical cycle of some elements are highly related to other elements. An example is the relationship existing between the iron, carbon, phosphorous, nitrogen and sulphur cycles (Madigan et al., 2008). Other unique microbial processes are mentioned in Table 1, even if this table does not exhaust the broad range of capabilities of microorganisms. The involvement of microorganisms and their diversity in the maintenance of natural ecosystems is of the most importance for understanding the functioning and conservation of National Parks.

Table 1. Characteristic microbial groups frequently detected in Doñana National Park and their representative metabolisms and living conditions.

Group	Characteristic metabolism	Environment
Domain Bacteria		
Gamma-Proteobacteria		
Pseudomonas	heterotrophic	Aerobe, facultative anaerobe
Enterobacteria	Heterotrophic Fe-chelated organic substrates	Facultative anaerobe
Delta-Proteobacteria		
Sulfate Reducing Bacteria	Sulfate reduction Sulfide production Organic carbon mineralization	Anaerobe
Thiobacillus	S oxidation NO_3^- reduction, N_2 production Denitrification	Anaerobe
Alpha-Proteobacteria		
Methylobacteriales	Methanotroph	Aerobe
Sphingomonadales	Heterotroph	Aerobe
Rhizobiales	Organic carbon N_2 fixation	Aerobe and microaerophile
Rhodobacter	Phototroph	Aerobe
Paracoccus	Denitrification	Aerobe
Epsilon-Proteobacteria		
Sullfurimonas	Sulfate oxidation, denitrification	Anaerobe
Firmicutes		
Bacillales	Heterotroph	Aerobe
Clostridiales	Litotroph, Heterotroph (a variety of metabolisms)	Anaerobe
Acidobacteria	Phototroph Heterotroph	Aerobe Aerobe/Anaerobe
Actinobacteria		
Actinomicetales	Heterotroph	Aerobe
Chloroflexi		
Roseiflexus	Heterotroph Photoheterotroph	Aerobe Anaerobe
Cyanobacteria	Phototroph N_2 fixation	Aerobe
Planctomycetes	Ammonium oxidation	Anaerobe
Bacteroidetes		
Bacteroidales	Heterotroph	Anaerobe
Flavobacteriales	Heterotroph	Facultative anaerobe
Domain Archaea		
Crenarchaeota	Amonium oxidizers	Aerobe
Methanogens	CH_4 production, acetate and H_2+CO_2 consumption	Anaerobe
Domain Eukarya		
Microalgae		
Chlorophyta Diatoms	Phototroph	Aerobe
Protozoa		
Flagellates Ciliates	Grazers of Bacteria and other microorganisms	Aerobe
Fungi	Heterothoph	Aerobe

Discussion

National Parks hold a number of unique environments, animals and plants. Today's landscapes are the result of millions of years of evolution where the tiniest and largest organisms have followed adaptative processes leading to the sites that are the target of preservation programs. At present, humans are a major cause altering the natural environments at an accelerated pace (Macpherson, 2007). A decrease of animal and plant diversity is easily visualized even if sometimes it is not reflected in visual environmental changes others that the lack of a unique species, often considered a natural symbol in the area. Examples could be represented by the trend to extinction of predators such as the lynx or imperial eagle in Doñana National Park (Mallinson, 1978; Ferrer and Calderon, 1990) which might not directly affect visible landscape changes.

Microorganisms are practically everywhere although the point to be underlined is how much diversity is maintained in specific natural environments (Nee, 2004; Curtis and Sloan, 2005). As seen in the presented results, microbial communities can be highly complex and their structure is not easily understood. What needs to be understood is that many microbial processes are only performed in nature by these tiny creatures. Thus, a diminishing microbial diversity will lead to the inhibition of some of these processes and consequently, an environmental degradation and transformation will occur (Horner-Devine et al., 2004). Thus, microbial diversity is essential for maintaining a healthy ecosystem. This implies that an ecosystem can easily and quickly respond to external factors altering the system equilibrium. A diversity of microorganisms permits a diversity of mechanisms to control, transform, or respond to external factors. A homogenization of microbial communities (Redford and Brosius, 2006) will be a first consequence of the lost of microbial diversity in nature. Low diversity in a system will lead to unresponsive systems which will be unable to recover from alterations in the ecosystem.

Unfortunately, human pollution and impact on pristine environments are being quite a common topic in current press (Peñuelas and Filella, 2002; Corsolini et al., 2002). Pollution, for instance, conducts to poor systems, where diversity decreases and so the capacity of the ecosystem to respond. Macroorganisms easily get into near extinction if their ecosystems get polluted above a certain threshold. Microorganisms constitute the most responsive force of the ecosystems. Able to perform a large variety of processes and armed by a large diversity of species, the microbial communities could be agents of biodegradation, for example, being capable of removing pollutants from a contaminated system (Larsson et al., 1988; Zaidi and Imam, 1999). The strength to achieve system stability provided by a large microbial diversity is not replaced by any other factor. As a consequence, the greatest potential to be preserved in any ecosystem is the microbial diversity which is responsible for maintaining the equilibrium of an environment. Diminishing microbial diversity might lead to a homogenization of microbial resources (Redford and Brosius, 2006) which contributes to increasing risks against the preservation of ecosystems in the long term. Thus, microbial diversity and function in natural ecosystems are aspects to be preserved if ecosystem responsiveness and stability needs to be maintained.

Microbial communities are complex and could represent the many pieces of a large puzzle. If these pieces are missing the system can hardly be completed. The capability of microorganisms to keep performing the large number of processes they currently carry out and others they potentially can perform depends on preserving intact the microbial world. Nevertheless, the microbial communities are rarely accounted for when biodiversity is considered (Colwell, 1997). In fact, at present, preserving microbial diversity is still difficult to achieve since we are still fighting to understand its level of complexity and functionality (Staley, 1997). Microbial diversity and the understanding of microbial community structure and its dynamism are goals for future research on the next years in order to consider that knowledge in conservation strategies for natural ecosystems, and especially for the National Parks.

Conclusion

Microbial diversity in Earth is huge and difficult to estimate. The importance of microbial diversity for ecosystem functioning is critical since numerous biogeochemical processes are exclusively carried out by microorganisms and potential changes in the environment get response from distinct members of the microbial community. As a consequence, preserving the large diversity of the microbial world is a necessary step to ensure long term stability and healthy environments, as well as achieving the goal in conserving their flora and fauna. This is the case for National Park conservation and the preservation of these unique environments.

Acknowledgements

This research has been supported by the Spanish Ministry of Education and Science (CGL2004-03927-C02-01/BOS) and the Andalusian Government (Bio288). The authors acknowledge the assistance of personnel from the Doñana National Park.

References

Altschul, S. F., Gish, W., Miller, W., Myers, E. W. & Lipman, D. J. (1990). Basic local lignment search tool. *Journal of Molecular Biology,* **215**, 403-410.

Colwell, R.R. (1997). Microbial diversity: the importance of exploration and conservation. *Journal of Industrial Microbiology and Biotechnology,* **18**, 302-307.

Corsolini, S., Romeo, T., Ademollo, N., Greco, S. & Focardi, S. (2002). POPs in key species of marine Antarctic ecosistem. *Microchemical Journal,* **73**, 187-193.

Curtis, T. P. & Sloan, W. T. (2005). Exploring microbial diversity – a vast below. *Science,* **309**, 1331-1333.

Curtis, T. P., Sloan, W. T. & Scannell, J.W. (2002). Estimating prokaryotic diversity and its limits. *Proceedings of the National Academy of Science USA,* **99**, 10494-10499.

Ferrer, M. & Calderon, J. (1990). The Spanish Imperial Eagle Aquila adalberti in Doñana National Park: a study of population dynamics. *Biology of Conservation,* **51**, 151-161.

Gonzalez, J. M., Ortiz-Martinez, A., Gonzalez-delValle, M. A., Laiz, L. & Saiz-Jimenez, C. (2003). An efficient strategy for screening large cloned libraries of amplified 16S rDNA sequences from complex environmental communities. *Journal of Microbiological Methods,* **55**, 459-463.

Gonzalez, J. M. & Saiz-Jimenez, C. (2004). Microbial diversity in biodeteriorated monuments as studied by denaturing gradient gel electrophoresis. *Journal of Separation Science,* **27,** 174-180.

Gonzalez, J. M., Portillo, M. C. & Saiz-Jimenez, C. (2005a). Multiple displacement amplification as a pre-PCR reaction to process difficult to amplify samples and low copy number sequences from natural environments. *Environmental Microbiology,* **7,** 1024-1028.

Gonzalez, J. M., Zimmermann, J. & Saiz-Jimenez, C. (2005b). Evaluating putative chimeric sequences from PCR amplified products and other cross-over events. *Bioinformatics,* **21,** 333-337.

Horner-Devine, M. C., Lage, M., Hughes, J.B. & Bohannan, B. J. M. (2004). A taxa-area relationship for bacteria. *Nature,* **432,** 750.

Larsson, P., Okla, L. & Tranvik, L. (1988) Microbial degradation of xenobiotic, aromatic pollutants in humic water. *Applied and Environmental Microbiology,* **54,** 1864-1867.

Macpherson, E. (2007). La protección y restauración de nuestros ecosistemas marinos, más allá de la creación de las áreas protegidas. *Ambienta,* June 2007, 40-46.

Madigan, M. T., Martinko, J. M., Dunlap, P.V. & Clark, D. P. (2008). *Brock Biology of Microorganisms* (12th Edn). San Francisco, CA, USA: Benjamin Cummings Publishing Co.

Mallinson, J. (1978). Lynxes. European lynx (*Lynx lynx*) and ardel lynx (*Lynx pardina*): *The shadow of Extintion.* London: MacMillan.

Nee, S. (2004). More than meets the eye. *Nature,* **429,** 804-805.

Pace, N. R. (1997). A molecular view of microbial diversity and the biosphere. *Science,* **276,** 734-740.

Peñuelas, J. & Filella, I. (2002) Metal pollution in Spanish terrestrial ecosystems during the twentieth century. *Chemosphere,* **46,** 501-505.

Portillo, M. C., Reina, M., Serrano, L., Saiz-Jimenez, C. & Gonzalez, J. M. (2008). Role of specific microbial communities on the bioavailability of iron in the Doñana National Park. *Environmental Geochemistry and Health,* **30,** 165-170.

Portillo, M. C. & Gonzalez, J. M. (2008). Statistical differences between molecular fingerprints from microbial communities. *Antonie van Leeuwenhoek,* **94,** 157-163.

Rappé, M. S. & Giovannoni, S. J. (2003). The uncultured microbial majority. *Annual Review Microbiology,* **57,** 369-394.

Rauch, E. M. & Bar-Yam, Y. (2004). Theory predicts the uneven distribution of genetic diversity within species. *Nature,* **431,** 449-452.

Redford, K. H. & Brosius, J. P. (2006). Diversity and homogenization in the endgame. *Global Environmental Change,* **16,** 317-319.

Serrano, L., Reina, M., Martín, G., Reyes, I., Arechederra, A., León, D. & Toja, J. (2006). The aquatic systems of Doñana. *Limnetica,* **25,** 11-32.

Siljeström, P. & Clemente, L. E. (1990). Geomorphology and soil evolution of a moving dune system in South-West Spain (Doñana National Park). *Journal of Arid Environments,* **18,** 139-159.

Staley, J. T. (1997). Biodiversity: are microbial species threatened? *Current Opinion in Biotechnology,* **8,** 340–345.

Ward, D. M., Weller, R. & Bateson, M. M. (1990). 16S rRNA sequences reveal numerous uncultured microorganisms in a natural community. *Nature,* **345,** 63-65.

Whitman, W. B., Coleman, D. C. & Wiebe, W.J. (1998). Prokaryotes: The unseen majority. *Proceedings of the National Academy of Science USA,* **95**, 6578-6583.

Wilcox, B. A. & Murphy, D.D. (1985). Conservation strategy: the effects of fragmentation on extinction. *American Naturalist,* **125**, 879-887.

Woese, C. (1987). Bacterial evolution. *Microbiological reviews* **51**, 221-271.

Zaidi, B. R. & Imam, S. H. (1999). Factors affecting microbiol degradation of polycyclic aromatic hydrocarbon phenanthrene in the Caribean coastal water. *Marine Pollution Bulletin,* **38**, 737-742.

Chapter 10

FIRE: A THREAT AND A LACKING TOOL FOR BIODIVERSITY CONSERVATION IN BRAZILIAN NATIONAL PARKS

L. Koproski[*1], *P.R. Mangini*[2] *and J.G. Goldammer*[3]

[1] UFPR – Universidade Federal do Paraná, Forest Sciences PhD Program.
[2] UFPR – Universidade Federal do Paraná, Environment and Development PhD Program
[3] Global Fire Monitoring Center (GFMC), Max Planck Institute for Chemistry, c/o Albert-Ludwigs–University Freiburg/United Nations University (UNU)

With more than 1.5 million species, Brazil is recognized as megadiversity country, hosting between 10 and 20% of the species already classified in the world. It is composed of six continental biomes, two of which are known as biodiversity hotspots: the Cerrado and the Atlantic Forest [1]. Nowadays, all Brazilian biomes are modified by human interventions, and their representative biodiversity is endangered at some level. To protect biodiversity in the last decades the number of protected areas in Brazil has grown rapidly. Currently there are nearly 300 federal protected areas officially established, with more than 24,000,000 hectares protected in 63 national parks. Worldwide fires in natural areas cause high environmental losses, negatively impacting global conservation efforts. The situation is not different in Brazil, where wildfires constitute a periodic occurrence and threat to some biomes, jeopardizing ecosystem functioning, biodiversity and atmosphere/climate-related processes. From 1979 to 2005, a total of 2,502 wildfires were officially registered in Federal Brazilian Protected Areas; 1,633 occurred in national parks. In 1979, only three occurrences were registered, contrasting with 196 occurrences in 2005 [2]. Even though these numbers should be higher, because many fire events are not registered or the records are improper, they are sufficient to represent the increasing number of fires in the last decades and to address fire as an important issue to biological diversity conservation in Brazilian National Parks.

---

*E-mail address: leticia@koproski.com

Fire is an agent with great ability to change environments. In some ecosystems fire can cause destruction, generate losses and irreparable ecological and economic damage. In other ecosystems, fire is essential for biodiversity variability due to the fact that most of the species have evolved in the presence of natural fires. In these ecosystems, fires may be necessary to maintain their original characteristics and can be used as an actively applied management tool to provide environmental health. In ecosystems where fire can be considered as a management tool for maintaining or restoring environmental balance, prescribed fires can be used to replace the benign impacts of natural fires. Prescribed fires could be effective means to shape disturbance-dependent ecosystem structures, to reduce fuels and thus wildfire intensity and severity and their effects on environmental health, which means a balance among ecosystems, vegetal, animal and human spheres of health [3]. As the Brazilian biomes have great variation from north to south, so do the fire regimes—the patterns of fire occurrence, size, severity, return intervals and ecological effects [4]—along the ecosystems.

In Brazil there are fire-sensitive ecosystems, such as the Amazon, and fire-dependent ecosystems, such as the Cerrado–Brazilian savannas. Fire regime conditions differ from intact to degraded within the same biome or eco-region. There are not very degraded fire regimes yet in Brazil; however, fire regime conditions are outside their natural range of variability, but are still considered restorable [5]. As fire impacts or benefits vary according environmental degradation status, at the present time, in Brazilian fragmented biomes conditions, even intact fire regimes could be, at some level, a biodiversity threat to degraded ecosystems. Thus, fire regimes should be established and applied to provide environmental health.

Considering fire-dependent environments, as the Cerrado biome and some grasslands and wetlands inside Atlantic Forest biome domain, Brazil has great potential and a widespread area, including national parks, which could be managed to restore fire regimes using prescribed fire. In the Cerrado domain alone, more than 2,000,000 hectares could be managed. North American, African and European countries have already put into practice prescribed burning in order to maintain and protect biodiversity inside protected areas, especially to maintain open and disturbance-dependent habitats. The Eurasian Fire in Nature Conservation Network (EFNCN) is a prominent example of its kind, providing advances in the use of prescribed fire in nature conservation, landscape management, forestry and carbon management in temperate-boreal Europe and adjoining countries in southeast Europe, the Caucasus, Central Asia and northeast Asia [6].

In 1989, the Brazilian Institute of Environment and Renewable Natural Resources (IBAMA) established the National System of Prevention and Control of Forest Fires (PREVFOGO) [7]. This system has been dealing with the fire issue and acting in protected areas. Despite the fact that the Brazilian National Park System presents a broad representation of different ecosystems, fire-dependent or not, in practice, the attempt to provide total fire exclusion still is, nowadays, the main method of fire management in protected areas of Brazil, although prescribed burning is predicted to be part of the national fire policy. According to experts in international organizations and research institutes, fire regimes in almost all ecosystems are threatened by inappropriate human introduction of fire, or inadequate fire suppression policies. In fire-dependent ecosystems, total fire suppression policies disturb fire regimes, leading to further degradation from fuel accumulation, loss of fire-adapted species, increases in fire-sensitive species, and extreme fire behavior when fires escape suppression forces.

As an example, in Paraná State, south of Brazil, fire represents a serious threat to biodiversity in Ilha Grande National Park. At the same time, it represents a tool that is lacking for biological diversity conservation. The Ilha Grande National Park, situated in the Atlantic Forest Biome, is located in the southern region of high Paraná river flood plain, in the boundary of the states of Paraná and Mato Grosso do Sul, between 23°16'/24°04' S and 53°43'/54°14' W. It covers 75,894 hectares of area, and 242,163 km of perimeter. It is composed by nearly 150 fluvial islands. The region is characterized by mesothermic humid subtropical climate (Cfa). The elevation varies from 200 to 220 m. Flooded grasslands are distributed for extensive areas, characterizing the main type of landscape in the park. The presence of 421 species of vertebrates, 52 of them mammals and 49 of them reptiles, had been registered in the area. Some of them are endemic and/or threatened species. The park has singular importance for biological diversity conservation, due to its influence by the aquatic cycles of the area, and its strategic geographic position as a biodiversity corridor [8]. Moreover, it still offers the scope and habitat diversity to support and maintain a wide range of species, including large threatened mammals such as marsh-deer (*Blastocerus dichotomus*), Brazilian tapir (*Tapirus terrestris*) and giant anteater (*Myrmecophaga tridactyla*). It also represents an important regional corridor to jaguars (*Panthera onca*) [9]. Wildfires occur frequently in the park; between 1999 and 2006, 88 events were registered. During this period, the total burned area was 208,234.6 hectares and human actions were the leading cause of fires in the park [10,11].

Although Ilha Grande National Park is located in Atlantic Forest domain, a fire-sensitive ecosystem, fire plays an essential role there in maintaining the environment structure due to the landscape and the main vegetation type. Besides human actions being the main cause of fires, lightning-ignited fires were registered too, and are a natural part of the ecosystem. Nowadays, total suppression is the park's fire policy. In the long term, sustaining this policy could be unfeasible, due to fuel accumulation and climatic dryness periods. The consequences of not successfully attaining this goal can be an additional threat to biodiversity, especially affecting the habitats of small vertebrates and even bigger mammals.

Between 2003 and 2005, more than 44 land vertebrates were observed during wildfires near the fire-line. Large mammals demonstrated behavioral adaptations to fire events, such as marsh-deer that usually remained closer to fire-line (Figure 1) without demonstrating disturbance, and the same behaviour was observed in tapirs, implying a better resistance to extreme fire events. In spite of large mammal resistance to those events, during extreme fire behaviour observed in the park, direct fire effects were observed on this group as well, as a marsh-deer recovered with burn lesions, and a sub-adult puma (*Puma concolor*) was found dead (Figure 2). However, the impacts of extreme wildfire on small and medium land vertebrates could be much more extensive, representing a serious threat to biodiversity considering the local vertebrates biomass and species variability included in those groups that are affected by these events. For example, the Pilosa order, formerly Xenarthra, are very sensitive to high-intensity and high-severity wildfires, and represented the most susceptible taxonomic group of medium-sized mammals. Several events registered demonstrate that nine-banded-armadillo (*Dasypus novemcinctus*) (Figure 3) and southern-tamandua (*Tamandua tetradactyla*) (Figure 4) were the species more affected by wildfire in the park, due to direct and indirect fire effects. Small vertebrates such as rodents, snakes and lizards were severely affected too, due to their reduced displacement capacity, crypt habits and inspected behaviours. Pit vipers (brazilian lancehead, *Bothrops moojeni*) (Figure 5) developed abnormal behavior near the flames, with static defensive and aggressive posture, increasing species vulnerability. In addition, extensive areas have been burned by wildfires every year in the park, changing food chain natural dynamics and forcing all sizes of land wildlife vertebrates to leave the park boundaries, driving animals to other threats such as road kill, exposure to domestic animal diseases, and illegal opportunistic hunting [10,12,13].

Figure 1 Marsh deer (*Blastocerus dichotomus*) near fire line during a wildfire in Ilha Grande National Park, Brazil.

Figure 2. Dead juvenile puma (*Puma concolor*) in Ilha Grande National Park, Brazil.

Figure 3. Carbonized nine-banded-armadillo (*Dasypus novemcinctus*) in Ilha Grande National Park, Brazil.

Figure 4. Dead southern-tamandua (*Tamandua tetradactyla*) in Ilha Grande National Park, Brazil.

Figure 5. Carbonized brazilian lancehead (*Bothrops moojeni*) dead due to fire effects in Ilha Grande National Park, Brazil.

The opposite example of fire impact over a preservation area, also in Paraná State, is the Cerrado State Park. There, fire should be considered a significant and multiple threat, not because it happens frequently, as in Ilha Grande, but in fact because events were not observed for more than 15 years even though it is situated in a fire-dependent ecosystem, leading to a hazardous fuel accumulation in the area. The park covers 430 hectares and is located in the eastern center of Paraná (24°10'S–49°39'W). The region is characterized by the mesothermic humid subtropical climate (Cfa) and the average altitude is about 810m. As the proper name of the unit suggests, it places itself in a classic Cerrado vegetation area, represented by its distinct physiognomies. The park shelters 40 mammals species, 270 bird, 45 reptile and 22 amphibian species. The Cerrado State Park represents the southern limit of this biome geographic distribution in Brazil, being the last remaining well-preserved area of this ecosystem in the State of Paraná [14]. If a wildfire happens in the park, it could be a catastrophic event, due to the fuel accumulation, and mainly due to the park's dimension. Its 430 hectares would probably be quickly consumed in an extreme fire event, which would certainly produce multiple impacts on local fauna and flora, carrying to expressive biodiversity losses. Besides the hazard of fuel accumulation, in an existing "efficient" long-term fire control and prevention program, the absence of fire in a fire-dependent ecosystem will drive some phytophysiognomies and their associated fauna to changes, modifying their original characteristics, which would go against the main goal of the unit—the conservation of the last remaining area of this original ecosystem in the State of Paraná.

In Brazilian fragmented biome conditions, a large burning extension is an extreme biodiversity threat to degraded ecosystems. If, in the past, animals and vegetation could recover from extensive fires, due to metapopulation dynamics nowadays this may not happen anymore. As national parks are becoming protected islands, without significant connections with other natural areas, it will be impossible for a variety of local wildlife to safely search for alternative areas during the post-fire period. Considering that fire is part of the ecosystem in parks as Ilha Grande and Cerrado State Park, as in many other areas in Brazil, this natural element of the ecosystem perhaps should be seen as a management tool, and have its policy control reviewed to restore and maintain environmental health. Prescribed fires could reduce hazardous fuel accumulation, and prevent extensive burning areas and extreme fire behaviour.

Fire management in Brazilian national parks could combine fire suppression for humans and even natural induced events as well as prescribed fires. The fire control polices should be planned to respect fire ecology in order to maintain fire in natural ecosystems, and at the same time should adequately consider the impacts upon biodiversity. Prescribed burnings must be conducted under weather and topographical considerations and a wise burning plan, established by considering various aspects such as burning size, type of fire, frequency, intensity, season of burning and fire history [15]. The planning should be defined according to the environmental requirements and protected area's purposes and objectives. This chapter intends to emphasize the possibility and the need to review actual polices on landscape management, considering the increased numbers and intensity of uncontrollable fire events in recent years in protected areas. The use of prescribed fire as a biodiversity conservation and fauna protection tool in national parks and other protected areas in Brazil could be one means to maintain targeted habitat structures, their representative biodiversity, and to reduce the destructive impacts of uncontrolled wildfires.

Acknowledgement

All pictures by Letícia Koproski.

References

[1] Mittermeier, R.A.; Myers, N.; Mittermeier, C.G.; Gil, P.R. *Hotspots, earth's biologically richest and most endangered terrestrial ecoregions.* Mexico: Agrupación Sierra Madre, 1999.

[2] PREVFOGO. Relatório de Ocorrência de Incêndios em Unidades de Conservação Federais 2005 [online]. 2008.11.20. Available at:_http://www.ibama.gov.br/prevfogo/ wp-content/files/roi_relatorio-2005.pdf

[3] Mangini, P.R. ; Silva, J.C.R. Capítulo 75: Medicina da Conservação: Aspectos Gerais. In: Cubas, Z.S.; Silva, J.C.R.; Catão-Dias, J.L. *Tratado de Animais Selvagens - Medicina Veterinária.* São Paulo: Roca LTDA, 2006, v.1, p.1258-1268.

[4] FAO. 2006. Fire management: voluntary guidelines. Principles and strategic actions. Fire Management Working Paper 17. [online]. 2008.11.20. Available at: http://www.fao.org/forestry/site/35853/en

[5] Shlisky, A.; Waugh, J.; Gonzalez, P.; Gonzalez, M.; Manta, M.; Santoso, H.; Alvarado, E.; Ainuddin Nuruddin, A.; Rodríguez-Trejo, D.A.; Swaty, R.; Schmidt, D.; Kaufmann, M.; Myers, R.; Alencar, A.; Kearns, F.; Johnson, D.; Smith, J.; Zollner, D.; Fulks, W.

Fire, Ecosystems and People: Threats and Strategies for Global Biodiversity Conservation. GFI Technical Report 2007-2. Arlington: The Nature Conservancy; 2007.

[6] EFNCN. Eurasian Fire in Nature Conservation Network. Available at: *http://www.fire.uni-freiburg.de/programmes/natcon/natcon.htm*

[7] IBAMA. Sistema Nacional de Prevenção e Combate aos Incêndios Florestais. [online]. 2008.11.20. Available at: *http://www.ibama.gov.br/prevfogo/*

[8] Campos, J.B. *Parque Nacional de Ilha Grande re-conquista e desafios*. 2 ed. Maringá: IAP/CORIPA, 2001.

[9] Mikich, S.B.; Bernils, R.S. *Livro vermelho da fauna ameaçada no Estado do Paraná*. Curitiba: IAP; 2004.

[10] Koproski, L.P. *O fogo e seus efeitos sobre a herpeto e a mastofauna no Parque Nacional de Ilha Grande (PR/MS), Brasil*. Curitiba, 2005. 127 p. Dissertation (Master's in Forest Engineer), Universidade Federal do Paraná.

[11] Koproski, L.; Nunes, J.R.; Soares, R.V.; Batista, A.C. *Improvement of FMA+ wildland fire danger índex to the Ilha Grande National Park (PR/MS), Brazil*. 4th International Wildland Fire Conference. Seville, 2007.

[12] Koproski, L.; Kuczach, A.M.; Mangini, P.R.; Pachaly, J.R.; Guerra-Neto, G.; Batista, A.C.; Kraus, D.; Goldammer, J.G. *Xenarthrans attitudes during wildfires in Brazil*. 6th International Zoo and Wildlife Research Conference on Behaviour, Physiology and Genetics. Berlin: Druckhaus Berlin Mitte GmbH, 2007.

[13] Koproski, L.; Mangini, P.R.; Pachaly. J.R.; Batista, A.C.; Soares, R.V. Impactos do fogo sobre serpentes (SQUAMATA) no Parque Nacional de Ilha Grande (PR/MS), Brasil. *Arq. Ciênc. Vet. Zool. Unipar*, Umuarama, v.9, n.2, p.129-133, 2006.

[14] IAP. Instituto Ambiental do Paraná. *Plano de manejo: Parque Estadual do Cerrado*. Curitiba: IAP; 2002.

[15] Goldammer, J.G; De Ronde, C. *Wildland Fire Management Handbook for Sub-Sahara África*. Global Fire Monitoring Center; 2004.

In: National Parks: Biodiversity, Conservation and Tourism   ISBN 978-1-60741-465-0
Editors: A. O'Reilly and D. Murphy, pp. 181-199   © 2010 Nova Science Publishers, Inc.

Chapter 11

VALUATION OF THE BENEFITS OF WILDLIFE TOURISM IN REMOTE PROTECTED AREAS: THE CASE OF GROS MORNE

Roberto Martínez-Espiñeira[*]
Economics, St. Francis Xavier University, PO Box 5000,
Antigonish B2G2W5, Nova Scotia, Canada.

Abstract

In this chapter the travel cost method is used to estimate the demand for wildlife viewing at a national park and to estimate consumer surplus measures associated with the access to the park for visitors mainly attracted to the park by the opportunities to observe wildlife. We use data collected on-site from Gros Morne National Park in Newfoundland&Labrador. Since the data were collected at the park, the analysis takes into account problems related to on-site sampling by using a negative binomial model. This model accounts for the non-negative and integer nature of the dependent variable and also corrects for the truncation and endogenous stratification in its distribution, as well as overdispersion. The results provide relevant insights for the management of recreational areas and for the preservation of biodiversity assets as an economic asset.

Keywords: wildlife viewing, moose, recreation demand, travel cost method, count data, endogenous stratification.

1. Introduction

The management of recreational sites such as national and provincial parks requires a balancing of both the costs and benefits associated with maintaining them. However, access to this type of recreational areas is often free or subject only to entry fees that clearly underestimate the maximum willingness to pay by most visitors, so their true value to the public is unknown and must be estimated using non-market valuation techniques. Perhaps the most common example of non-market valuation technique applied to the valuation of

---

[*] E-mail address: rmespi@stfx.ca, Tel: 1-902-867-5433, Fax: 1-902-867-3610.

recreational sites is the travel cost method. Using this method, the benefit derived from access to a recreational site can be estimated by observing the trade-off between the cost of reaching the site and the numbers of visits made. In single-site travel cost studies, the number of visits to the site is assumed to be related to travel cost, time, and other demographic or locational variables (Parsons, 2003). This allows the researcher to estimate a trip demand curve and from it calculate welfare measures associated with having access to the site (for example, consumer surplus). By restricting this observation to those users who stated that the main purpose of their visit was related to a particular type of recreational activity, one can also evaluate the value of the site for that type of user.

In this chapter, the focus is the estimation of the value of Gros Morne National Park for non-consumptive wildlife recreation. To this end, those whose decision to visit was strongly influenced by the possibility of enjoying opportunities to encounter wildlife are isolated from the main sample of visitors to the park. In the case of Gros Morne National Park, which harbors a large number of highly visible wildlife species, it is expected that the benefit derived from accessing the park by visitors interested in wildlife-based recreation is larger than for the average visitor.

With an on-site sample, only those visitors who have made at least one visit are observed, so the distribution of the variable visits is truncated at zero. Additionally, since visitors were intercepted on site, the probability that a visitor has been intercepted is positively related to the number of visits made, leading to an issue of endogenous stratification of the variable visits. The econometric analysis used below accounts for these features of the distribution of the dependent variable used to estimate the demand functions for wildlife-based recreation.

In the next section, we present a brief description of the Travel Cost Method followed, in Section 3., by the methodology of the survey and the data collection procedures. Section 4. deals with the econometric and estimation issues. The data description and the choice of variables for the estimated model appear in Section 5.. Section 6. includes the discussion of the estimation results, followed by the conclusions.

2. The Travel Cost Method

The Travel Cost Method (TCM) is one of several revealed preference methods applied to the valuation of non-marketed goods and services (Braden and Kolstad, 1991; Freeman 1993, Garrod and Willis 1999) and it is particularly well-suited to the valuation of recreational sites and wildlife sanctuaries. The method assumes that even if access to a site has only a minimal or null explicit price, visitors' travel costs (including transportation, accommodation, and lost wages) may be used as a proxy for the relevant price, assuming that visitors perceive and react to changes in travel costs as they would to changes in an entry fee. Under these assumptions, the number of trips to a recreation site is expected to decrease in distance traveled and other factors that increase the total travel cost. A demand relationship can be then estimated by exploiting this relationship between the number of visits and the size of travel costs. Additional arguments, such as socioeconomic characteristics of the visitors and information concerning substitute sites and environmental quality indicators, can also be included in the demand function.

It should be noted that it is the weak separability of recreation demand from non-recreation consumption and the weak complementarity (Mäler, 1974) of the marketed goods and services required to get to and to enjoy the site that make it possible to estimate a demand curve for individual sites and, from it, a measure of the consumer surplus derived from the site. Therefore, the TCM can measure only user values derived from the recreational site. The TCM cannot calculate any type of non-use value (Krutilla, 1967), such as intrinsic value, existence value, option value, or bequest value. This is particularly relevant in the case of the valuation of wildlife resources associated with the site, since non-use values derived from many wildlife species can be substantial. The estimates of full-economic value obtained from TCM studies will therefore err on the conservative side and can only be considered as a lower-bound measure of the full benefit of recreational sites.

The TCM has been previously used in several instances to estimate the recreational value of wildlife resources and/or to estimated the demand for wildlife by tourists. Some of these studies are Cooper and Loomis (1991); Bayless, Bergstrom, Messonnier, and Cordell (1994); Zawacki and Bowker (2000); Marsinko, Zawacki, and Bowker (2002); and Gürlük and Rehber (2008).

3. Data

Gros Morne National Park covers 1,805 Km^2 on the Southwestern side of the Great Northern Peninsula in the Canadian province of Newfoundland and Labrador. The park was established in 1973 and was also identified in 1987 as a UNESCO World Heritage Site, due to its rather unique geological features (in particular the Tablelands, the Long Range Mountains and Western Brook Pond). It is considered one of Canada's most spectacular and unspoiled locations. In fact, Gros Morne is a key contributor to the Newfoundland's appeal as an exotic, high quality wilderness area (Locke & Lintner, 1997). Newfoundland and Gros Morne offer great opportunities for wildlife recreation not only because it is relatively easy to experience encounters with wildlife but also because long isolation has produced a large number of Newfoundland subspecies: nine of the fourteen land mammals native to the Island are distinct from their mainland relatives.[1]

About 28% of the park is forested and the forest is composed primarily of balsam fir and white birch, and to a lesser extent white and black spruce. Most visitors hike in the park mainly during the peak season of July and August. Their hiking experience provided by the varied and attractive scenery is also enhanced by the abundant opportunities to encounter wildlife. Some of the species that can be found in the park include arctic hare, black bear, beaver, deer, caribou, and moose. Moose, which were introduced to Newfoundland in 1904, are now particularly abundant in Gros Morne. In Gros Morne National Park, there are approximately 7,800 moose with an average density of 4.3/km2. In some locations, the density is as high as 19.5/km2,[2] making it one of the highest ever recorded in North America. Therefore, the opportunities for viewing moose constitute for some visitors one of the main reasons to visit the area. Other popular recreational activities that include angling, swimming, and whale watching contribute to attracting approximately 120,000 visitors to

---

[1] See http://www.pc.gc.ca/pn-np/nl/grosmorne/natcul/natcul3_E.asp
[2] See http://www.pc.gc.ca/apprendre-learn/prof/itm3-guides/vraie-true/etu-stuplan3case2_e.asp.

the park annually.

The data used in this study were collected through an on-site survey of visitors conducted between June and September 2004. Using the 'next available vehicle' methodology at the entries of the park and intercepting on-foot visitors at a series of hotspots within the park, a team of interviewers randomly sampled visitors daily (except Sundays). Interviewers were distributed across the park according to a careful sampling plan (developed by Parks Canada) ensuring that visitors from all origins and using different facilities had the same likelihood of being interviewed. Visitors were briefly interviewed (mainly about party size and region of residence), given a questionnaire, and asked to fill it in and mail it back after their visit to the Park. A total of 3140 questionnaires were administered with 1213 returned, giving a response rate of 0.386. Note that the format of the survey prevented the use of reminders, since interviewers only asked about zipcodes and postcodes, rather than actual names and addresses.

The questionnaire included among others questions on the main reasons for the trip, number of times the respondent had visited the park in the previous five years, home location, duration of visit, attractions visited, income, travel cost, size and age composition of travel party, distance to substitute sites, and other sites visited on the same holiday. For further details about the survey effort, the questionnaire, and the data obtained see Parks Canada (2004a, and 2004b) and D. W. Knight Associates (2005).

4. Econometric Methods

Given the characteristics of the variable that reflects the number of visits to a site, count data models are the most commonly used to estimate single-site recreation demand models (Creel and Loomis, 1990; Englin & Shonkwiler 1995a; Gurmu & Trivedi, 1996; Shrestha, Seidl, and Moraes, 2002).[3] Regression models for counts account for the fact that the response variable is discrete and takes nonnegative integer values only. Count data distributions are also characterized by a concentration of values on a few small discrete values (such as 0, 1 and 2), skewness to the left, and often show intrinsic heteroskedasticity with variance increasing with the mean (Cameron & Trivedi 1998 and Cameron, 2001).

4.1. Poisson

The most basic count models are based on the Poisson distribution. Hellerstein and Mendelsohn (1993) provide a theoretical basis for the use of count data to model recreational demand. On any given choice occasion, the individual's decision about whether to take a trip or not can be modeled with a binomial distribution. As the number of choices increases this asymptotically converges to a Poisson distribution. The density of this distribution for the count (y) is given by:

$$\Pr[Y = y] = \frac{e^{-\mu}\mu^y}{y!}, \qquad y = 0, 1, 2, \ldots \qquad (1)$$

---

[3] See Englin, Holmes, and Sills (2003) for a summary of the history of the application of count data models to recreation demand analysis.

where μ is the visitation intensity or rate parameter. The mean and the average of this distribution equal each other ($E[Y] = \mu = V[Y]$), a property known as *equidispersion*. This model can be extended to a regression framework by parameterizing the relation between the mean parameter μ and a set of regressors x. An exponential mean parametrization is commonly used:

$$\mu_i = \exp(x'\beta), \qquad i = 1, ..., n \qquad (2)$$

where x is the matrix of k regressors and β is a conformable matrix of coefficients to be estimated. Since $V[y_i|x_i] = exp(x'_i\beta)$, the Poisson regression is intrinsically heteroskedastic. Given (1) and (2), the Poisson regression model can be estimated, under the assumption that $(y_i|x_i)$ are independent, by maximum likelihood.

4.2. Negative Binomial

In practice, however, data on the observed number of visits to a recreational site show overdispersion relative to the Poisson distribution. That is, the variance of the number of visits is larger than the mean, because a few respondents make a large number of visits while most respondents make only a few. This makes the use of the Poisson model overly restrictive. Overdispersion has qualitatively similar consequences to heteroskedasticity in the linear regression model. Therefore, as long as the conditional mean is correctly specified, the Poisson maximum likelihood estimator with overdispersion is still consistent, but it underestimates the standard errors and inflates the t-statistics in the usual maximum-likelihood output. In intuitive terms, this would have the effect of making the influence of the explanatory variables in the model look artifically large (Hilbe, 2007; Martínez-Espiñeira & Hilbe 2008).

When the overdispersion problem is serious, a widely-used alternative is the negative binomial model. This is commonly obtained by introducing an additional parameter that reflects the unobserved heterogeneity that the Poisson fails to capture. Let the distribution of a random count y be Poisson, conditional on the parameter λ, so that $f(y|\lambda) = exp(-\lambda)\lambda y/y!$ and assume that the parameter λ is random, rather than being a completely deterministic function of the regressors x. In particular, let $\lambda = \mu\nu$, where μ is a deterministic function of x, say $\mu = exp(x'\beta)$, letting $\nu > 0$ be independently and identically distributed with density $g(\nu|\alpha)$, where α is denoted the overdispersion parameter. This is an example of unobserved heterogeneity, as different observations may have different λ (heterogeneity) but part of this difference is due to a random (unobserved) component ν, which the Poisson regression model would fail to capture.

If $f(y|\lambda)$ is the Poisson density and $g(\nu)$, $\nu > 0$, is assumed to be the gamma density with $E[\nu] = 1$ and $V[\nu] = \alpha$, we obtain the negative binomial density:

$$h(y|\mu, \alpha) = \frac{\Gamma(\alpha^{-1} + y)}{\Gamma(\alpha^{-1})\Gamma(y + 1)} \left(\frac{\alpha^{-1}}{\alpha^{-1} + \mu}\right)^{\alpha^{-1}} \left(\frac{\mu}{\mu + \alpha^{-1}}\right)^y \qquad \alpha > 0 \qquad (3)$$

where $\Gamma(\cdot)$ is the gamma function. The parameter α determines the degree of dispersion in the predictions. Special cases of the negative binomial include the Poisson ($\alpha = 0$) and the geometric ($\alpha = 1$). A likelihood-ratio test based on the parameter α can be employed to test the hypothesis of no overdispersion.[4]

4.3. Truncation

When the data are collected on site, the distribution of the response variable is also truncated at zero, since only those visitors who made at least one visit are observed. If this truncation is not accounted for the resulting estimates will be biased and inconsistent, because the conditional mean is misspecified (Shaw, 1988; Creel & Loomis, 1990; Grogger and Carson, 1991; Yen & Adamowicz, 1993; Englin & Shonkwiler, 1995a). The density of the Poisson distribution truncated at zero for the count (y) is given by:

$$\Pr[Y = y|Y > 0] = \frac{e^{-\mu}\mu^y}{y!} \cdot \left[\frac{1}{1 - e^{-\mu}}\right], \qquad y = 1, 2, \ldots \qquad (4)$$

The basic Poisson model is unbiased even under overdispersion but this is not the case with the truncated version of Poisson. If there is significant overdispersion, the truncated Poisson model yields inconsistent and biased estimates (Grogger & Carson, 1991). In that case, the truncated negative binomial should be used instead. The density of the negative binomial distribution truncated at zero for the count (y) is given by:

$$\Pr[Y = y|Y > 0] = \frac{\Gamma(y + \alpha^{-1})}{\Gamma(y + 1)\Gamma(\alpha^{-1})}(\alpha\mu)^y(1+\alpha\mu)^{-(y+\alpha^{-1})} \cdot \left[\frac{1}{1 - (1 + \alpha\mu)^{-\alpha^{-1}}}\right] \qquad (5)$$

Examples of applications of this model in the travel cost literature include Bowker, English, and Donovan (1996); Liston-Heyes and Heyes (1999); Zawacki and Bowker (2000); and Shrestha et al. (2002), while Yen and Adamowicz (1993) compare welfare measures obtained from truncated and untruncated regressions.

4.4. Endogenous Stratification

One last issue to consider is that, since the data were collected on-site, the distribution of the response variable on the number of visits is endogenously stratified. This is because the visitors' likelihood of being sampled is positively related to the number of trips they made to the site: frequent visitors are more likely to be sampled. This issue was first addressed by Shaw (1988), while Englin and Shonkwiler (1995a) extended their analysis with an application of the truncated and endogenously stratified negative binomial model. Dobbs (1993) considered the problem in a continuous context, while Moeltner and Shonkwiler (2005) extended the correction to the case of count data random utility models.

If the assumption of equidispersion holds, standard regression packages can be used to estimated a Poisson model adjusted for both truncation and endogenous stratification. In this case Shaw (1988) shows that:

$$\Pr[Y = y|Y > 0] = \frac{e^{-\mu}\mu^{y-1}}{(y-1)!}, \qquad y = 1, 2, \ldots \qquad (6)$$

so it suffices with regressing $y^* = y - 1$ with a conventional Poisson regression model (Haab & McConnell, 2002, p. 174-181).

---
[4] See Cameron and Trivedi (1990) or Cameron and Trivedi (2001, p. 336) for details.

This very convenient model has been used in several applied studies (Bin, Landry, Ellis, and Vogelsong 2005; Hagerty & Moeltner, 2005) under the assumption that overdispersion is not significant (Fix & Loomis, 1997; Hesseln, Loomis, González-Cabán, and Alexander., 2003; Loomis, 2003).

For the case where the equidispersion assumption is not tenable, the density of the negative binomial distribution truncated at zero and adjusted for endogenous stratification for the count (y) was derived by Englin and Shonkwiler (1995a) as:

$$\Pr[Y = y|Y > 0] = \frac{\Gamma(y_i + \alpha_i^{-1})}{\Gamma(y_i + 1)\Gamma(\alpha_i^{-1})} \alpha_i^{y_i} \mu_i^{y_i-1} (1 + \alpha_i \mu_i)^{-(y_i + \alpha_i^{-1})} \qquad (7)$$

Contrary to the equidispersed case above, this expression cannot be rearranged into an easily estimable form. Therefore, until recently (Hilbe & Martínez-Espiñeira 2005) it could not be easily handled with basic routines included in econometric software packages, requiring instead to be programmed as a maximum likelihood routine, with the associated increase in computational burden. Englin and Shonkwiler (1995a); Curtis (2002); Englin et al. (2003); Ovaskainen, Mikkola, and Pouta (2001); Martínez-Espiñeira and Amoako-Tuffour (2008) used variants of this model.

It should be noted that this last model accounts simultaneously for the thereе features of the response variable when the data were collected on-site: zero-truncation, overdispersion, and endogenous stratification. Martínez-Espiñeira, Amoako-Tuffour, and Hilbe (2006); Martínez-Espiñeira, Loomis, Amoako-Tuffour, and M.Hilbe (2008); and Martínez-Espiñeira and Hilbe (2008) show that in empirical applications of the travel cost method, most of the correction needed to adjust welfare measures towards their correct values tends to be associated with the issues of overdispersion and truncation.

Often, the empirical applications of this model are based on a variation of Equation (7) that restricts α to a common value for all observations (so $\alpha_i = \alpha$). However Englin and Shonkwiler (1995a) parameterized α as $\alpha_i = \frac{\alpha_0}{\lambda_i}$. Ovaskainen et al. (2001) did also try this specification but found better results by fixing α for all observations at a value previously estimated using a nonlinear squares regression. McKean, Johnson, and Taylor (2003) allowed α to vary as a function of a randomly generated parameter, but that parameter was not related to visitor characteristics.

In this chapter we use the more flexible approach that allows the overdispersion parameter to vary according to the characteristics of the visitor and compare it with the more restrictive approach. The analysis was conducted in STATA 9.1 using the commands $nbstrat$ (Hilbe & Martínez-Espiñeira, 2005) and $gnbstrat$ (Hilbe, 2005).

5. Model Specification

The analysis parts from the definition of a simple single-site demand function with the general form:

$$Y_i = f(TC_i, S_i, D_i, I_i, V_i) \qquad (8)$$

where TC_i is the travel cost, inclusive of the cost of travel time, which acts as the price variable in the demand function; S_i stands for information on substitutes sites; D_i

represents socio-demographic characteristics of the respondent and the visitor party; I_i is a measure of income; V_i are features of the current visit to the park; and i indexes the individual respondents.

The dependent variable of this demand function is in this case *persontrip*, which is constructed as the product of *partysize* during the current trip times the number of times (including the current trip) the respondent visited Gros Morne during the past five years. By using this variable as opposed to the number of visits during the period, the problem of lack-of-dispersion, endemic to individual Travel Cost Method models (Ward & Loomis, 1986), can be circumvented (Bowker et al., 1996)). Bhat (2003) also used this format for the Florida Keys because, as in the case of Gros Morne, group travel by car is very common in the Florida Keys (Leeworthy & Bowker, 1997), while Heberling and Templeton (2008) used the same type of variable to their study of trips to the Great Sand Dunes National Park and Preserve. The calculation of *persontrip* makes the simplifying assumption that the same number of people formed the partysize in all the trips during the previous five years. This type of assumption is implicitly made in travel cost method studies of this type about all the variables built by extrapolating information from the characteristics of the last trip made. Asking respondents to provide detailed differentiated information about each of the previous trips is usually considered as imposing excessive recall burden.

The independent variables in Expression 8 were also constructed on the basis of answers to the four-page, 27-question questionnaire (whose full-text is available upon request). From the home location responses (postcode for Canadian residents, zipcode for US residents, and country for residents of other countries) flying distances to the airport of entry or driving distances to Gros Morne (measured as driving distances to Rocky Harbour the town located in the center of the Park) were calculated.

The construction of the price proxy $Travelcost$ (measured in CAN$ 1000 per year) follows a generic accounting approach commonly adopted in the travel cost literature (Hesseln et al., 2003; Englin et al. 2003). The travel cost is calculated as the product of the number of round-trip kilometers from the visitor's residence to Gros Morne times 0.35 $CAN/Km if the visitor entered Newfoundland by ferry (and therefore by car). However, If the visitor entered through any of the airports in Newfoundland, the assumption is made that she took a flight from her residence and the cost of flying is estimated at $CAN/Km 0.20 for one-way distances under 4000 Km and 0.10 $CAN/Km for one-way distances over 4000 Km. This simplifying approximation to the real out-of-pocket expenses associated with transportation costs is similar to the one used by Bhat (2003). Only information on the visitor's point of entry into Newfoundland was available, not on the modes of transportation used during the whole trip. From this information, we had to infer the mode of transportation used for the whole trip. However, it is likely that some visitors flew from their home to, say, Halifax in Nova Scotia or one of the main hubs in central Canada (Montreal, Toronto, or Ottawa) before renting a car to continue the rest of their trip to the park. The travel cost for those visitors was calculated still assuming that they had driven all the way from home to the park. However, one could expect that, since the daily cost of renting a car exceeds the opportunity cost of using a private car, the finally calculated driving cost per kilometer should average to more or less the same as the cost per kilometer worked out assuming that they just drove all the way from home in their own car. Those visitors who entered Newfoundland through the international airport in St. John's (located on the East Coast of

Newfoundland) had to drive 700 Km to Gros Morne. This was accounted for when calculating *travelcost* too. This variable is then divided by *partysize* before adding it to the estimated cost of travel time (see definition of *traveltimecost* below) to compute the full cost of traveling to the park (Cesario, 1976). Some studies (e. g. Fix & Loomis, 1998) use reported travel costs. In order to avoid survey overburden and minimize response and recall bias, respondents were not asked to calculate their travel costs themselves. Bowker et al. (1996) report models using both approaches to variable travel cost calculation, finding no appreciable differences.

The valuation of travel time (*traveltimecost* in our notation) remains a thorny problem for practitioners of the travel cost method (Englin & Shonkwiler, 1995b; Feather & Shaw, 1999; Zawacki & Bowker, 2000; Hesseln et al., 2003; McKean et al. 2003; Amoako-Tuffour and Martínez-Espiñeira, 2008). Amoako-Tuffour & Martínez-Espiñeira (2008) explicitly consider the problem of estimating the opportunity cost of time in the case of visits to Gros Morne. In this chapter, however, a simple approach was adopted based on using the product of round trip time times 0.3 of the wage rate to proxy the opportunity cost of time. Similarly, Cesario (1976) used the same approach with 0.43 as the relevant fraction, Zawacki and Bowker (2000) and Bowker et al. (1996) use 0, 0.25, and 0.5 instead, while Liston-Heyes and Heyes (1999) and Hagerty and Moeltner (2005) use 1/3. Sohngen, Lichtkoppler, and Bielen (2000) and Sarker and Surry (1998) use 0.3. The wage rate was roughly approximated as the ratio of the annual income divided by 1880 hours of work per annum (Sohngen et al. , 2000; Bin et al., 2005). Travel time was estimated from the estimated travel distance to the Park by assuming a driving average speed of 80 Km/hour and a flying average speed (for those whose point of entry was one of Newfoundland's airports.) of 600 Km/hour. Due to the high collinearity between the two measures, it was impossible to include them separately in the model. Multicollinearity may result in wrong signs and/or implausible results (Smith, Desvouges & McGivney, 1983; Englin & Shonkwiler, 1995; Earnhart 2004). Instead, the cost of travel time *traveltimecost* and the cost of travel *travelcost* were combined into CTC: the proxy of the travel cost that acted as a price in Expression 8. This price proxy was measured in CAN$ 1000.

The expected effect of the time spent at the park ($daysatGM$): was uncertain a priori, although Shrestha et al. (2002) and Creel and Loomis (1990) find that the longer the duration of the trip the less the trips taken and Bell and Leeworthy (1990) also find that people living far away make fewer trips but stay longer at the site.

If a visitor lives near a substitute recreational site which affords opportunities for the type of recreational activity conducted at the valued site, in this case observing wildlife, the number of trips to the site analyzed will likely decrease. Rather than including the distance to a stated substitute site, we followed Bowker et al. (1996) and used a dummy (*substitute*) that takes the value of one if the respondent suggested an alternative site or the distance to it. Liston-Heyes and Heyes (1999) chose not to include the distance to substitute sites and describe the difficulties involved in introducing this variable in the demand model. McKean et al. (2003) also found the effect of this variable non-significant and many respondents failing to provide a value for it.

Respondents were also asked if they had visited other national parks in the Atlantic region, as in Liston-Heyes and Heyes (1999). The final model included dummies for Terra Nova National Park ($TerraNova$) and Cape Breton Highlands ($Highlands$) National Parks.

The questionnaire included questions about the visitor's reasons for visiting Newfoundland and Labrador and the relative influence of Gros Morne in the decision to visit this province: This made it possible to screen out those visitors from outside the province whose decision to visit Newfoundland and Labrador had little to do with their visit to Gros Morne. Similar variables were also used by Beal (1995); Sohngen (1998); and Liston-Heyes and Heyes (1999). A more comprehensive treatment of multi-site and multi-purpose visitors to Gros Morne is offered by Martínez-Espiñeira and Amoako-Tuffour (2008).

Other variables considered in the model were the number of people in the visitor group sharing expenses in the current trip ($partysize$) as in Liston-Heyes and Heyes (1999) and (2003) (2003) and age composition of the visitor group in the current trip (Siderelis and Moore 1995). The proportion of party members under seventeen ($propou17$) was used in the final models.

The effect of visitors' *income* (which we measured in $CAN1000) is often found to be weak in travel cost studies. Many studies found it negative or non-significant (Creel and Loomis, 1990; Sohngen et al., 2000; Loomis, 2003). Liston-Heyes and Heyes (1999) find visits an inferior good. Bin et al. (2005) find a significant positive effect of income on the number of trips to North Carolina Beaches. Given the remoteness of Gros Morne, we expected income to exert a positive effect on the number of visits, even though residents of Newfoundland, whose average income is relatively low, would have of course visited very often. Similarly, the level of educational attainment ($educat$) was expected to show a positive effect a priori, although Shrestha et al. (2002) find a negative effect on fishing trips

Visitors were asked about the degree of influence of several activities, in addition to the prospect of encountering wildlife, such as hiking, backpacking, and camping and different features (such as the fact that it is a World Heritage site) of the park in their decision to make the visit. The variable $camping$ (about the influence of camping) was kept in the final model to explain the overdispersion parameter α.

6. Results

In order to focus on those visitors for whom wildlife was a strong influence in their decision to visit Gros Morne, we eliminated all visitors who, on a scale from 0 to 10, stated that the prospect of encountering wildlife had an influence of 7 or less. No other visitors were eliminated from the sample. In particular, note that long haul travelers, which are probably more likely to visit the park as part of a multipurpose or multi-site trip, were not eliminated from the sample. This is because addressing the issues associated with multi-purpose and multi-site trips is beyond the scope of this chapter. The interested reader is directed to Martínez-Espiñeira and Amoako-Tuffour (2008) for a detailed account of these problems and how they have been addressed in the TCM literature, as well as an application of one of the proposed methods to address the issue to the case of the same sample used here.

Table 2 shows the results of two econometric specifications that account not only for the truncation and overdispersion in the distribution of the dependent variable, but also for the endogenous stratification resulting from the fact that an on-site sampling was used. Model

Table 1. Summary descriptives of the variables used in the econometric model (N=414)

Variable		Mean	Std. Dev.	Min	Max
camping	influence of camping facilities	3.71	3.86	0	10
CTC	combined travel cost	1.46	1.29	0.01	8.85
daysatGM	days spent at Gros Morne	3.80	2.54	0.5	21
educat	education level	4.15	1.08	1	6
expenses	estimaated expenses while at Gros Morne	0.24	0.22	0	1.5
flew	= 1 if respondent flew into Newfoundland	0.38	0.49	0	1
Highlands	= 1 if respondent visited Highlands during current trip	0.30	0.46	0	1
income	income level of respondent	88.05	43.63	20	160
missexp	= 1 if expenses value was missing	0.08	0.28	0	1
missincome	= 1 if income value was missing	0.08	0.27	0	1
partysize	number of visitors sharing expenses	2.58	1.25	1	10
persontrip	number of visits in last 5 years times partysize	3.70	6.16	1	84
prop3465	proportion of the visiting party aged 35 to 65	0.63	0.41	0	1
propu17	proportion of the visiting party under 17	0.06	0.16	0	1
substitute	=1 if viistor provided a substitute to Gros Morne	0.64	0.48	0	1
TerraNova	= 1 if respondent visited Terra Nova during current trip	0.31	0.46	0	1

$GNBSTRAT$ also allows the overdispersion parameter α to vary across visitors according to characteristics of the visitor group.

The results from models based on the Poisson distribution are not reported, since the overdispersion problem appeared significant, so these models would be overly restrictive. In fact, a likelihood-ratio test of $\alpha = 0$ based on comparing the zero-truncated Poisson versus the Negative Binomial results in a $\overline{\chi}^2(01) = 456.99$ with $Prob >= \overline{\chi}^2 = 0.000$. The equivalent likelihood-ratio test between a plain negative binomial and a plain Poisson (not reported but available upon request) yields $\overline{\chi}^2(01) = 401.85$ with $Prob >= \overline{\chi}^2 = 0.000$.

It can be seen that under $GNBSTRAT$ the log-likelihood improves and the size of the coefficient of the price proxy variable (CTC) also rises as compared to $NBSTRAT$. The latter means that the consumer surplus measures become larger. $GNBSTRAT$ dominates $NBSTRAT$ by allowing the overdispersion parameter α to vary according to $income$ and other variables describing the visitor group. A Likelihood-ratio test yields the value of $\chi^2(2) = 227.59$ with $Prob > \chi^2 = 0.0000$. The coefficients of all the covariates in the α equation are highly significant, confirming that using the same overdispersion parameter for all observations would be overrestrictive.

At the bottom of Table 2 pseudo-R^2 values are reported. These are calculated as $R^2 = 1 - \frac{\ln L}{\ln L_0}$ where $\ln L$ is the log-likelihood of the full model and $\ln L_0$ is the log-likelihood of a model with the restriction $\beta = 0$. This measure of fit is not comparable to ordinary least squares R^2, but still provides an indication of the improvement of the fit of the model over a restricted model with only a constant term. Following Ovaskainen et al. (2001) and Martínez-Espiñeira and Amoako-Tuffour (2008), for a common point of reference and for the overall rather than incremental fit for the negative binomial, the restricted log-likelihood of the relevant Poisson (with restrictions $\beta = 0$, $\alpha = 0$) was used as the restricted log-likelihood for both the Poisson and negative binomial models when computing the pseudo-R^2 so both $NBSTRAT$ and $GNBSTRAT$ are each compared to a restricted model of a zero truncated Poisson that accounts for endogenous stratification. This measure confirms that the most flexible model works best. The discussion below therefore corresponds to Model *GNBSTRAT*.

Table 2. Estimation results of NBSTRAT and GNBSTRAT count data models. Dependent Variable is *persontrip*.

Variable	NBSTRAT	GNBSTRAT
CTC	-0.4372***	-0.2449***
income	-0.0021	0.0060***
substitute	0.5191***	0.1645*
educat	-0.0324	-0.0349
expenses	-1.7541***	-0.7871**
daysatGM	0.1138***	0.0750***
missincome	-0.4313***	-0.1064
missexp	0.4501	0.3079*
Highlands	0.0414	-0.0393
TerraNova	-0.4290***	-0.2006**
constant	0.2512	0.0155
$\ln(\alpha)$		
income		-0.0236***
camping		0.1719***
prop3465		2.1318***
propu17		2.5783***
partysize		0.7704***
flew		-1.2583***
constant	10,010	-2.8477***
Statistics		
log-likelihood	-866.9	-753.1
pseudo R^2	0.37	0.45
Observations	414	414
χ^2	92.73	68.72
AIC	4.241	3.691
legend: * p<0.1; ** p<0.05; *** p<0.01		

The variable used to approximate the effective price of a trip (CTC) presents the expected negative sign, which yields a negatively sloped demand curve for *persontrips*. This means that the further away a visitor lives the fewer the visits to the park in the past five years and/or the smaller the visitor party in the current trip. The effect of the variable *income* appears significant and has a positive sign only when the overdispersion parameter is allowed to vary across visitors. Often income is found to be non-significant in travel cost studies, but it is likely that the remote location of Gros Morne makes the visit expensive enough for many visitors for visits to be a normal good. Bin et al. (2005) find a significant positive effect of income on the number of trips to North Carolina Beaches. The variable *educat* presents a negative but non-significant sign. The variable *expenses* presents the expected negative sign: those who tend to spend more on a visit to the park, tend to make fewer trips. This can also be explained by the fact that frequent visitors will have invested in equipment (particularly camping equipment) that makes their on-site expenses lower.

The length of the visit ($daysatGM$) has a significant and positive sign. Bowker et al. (1996) also find a positive sign for time spent at the site. However, this result is at odds with previous findings. Shrestha et al. (2002) and Creel and Loomis (1990) find that the longer the duration of the trip the fewer the trips taken and Bell and Leeworthy (1990) also find that people living far away make fewer trips but longer stays. The fact that the length of stay appears positively correlated with the frequency of visits may be associated with the remote geographical location of Gros Morne and the type of recreational activities that it offers.

The binary variable *substitute* has a significant positive sign. In theory, we would have expected that those visitors who came up with a next best alternative to Gros Morne would visit this park less frequently (e.g. Parsons, 2003). However, it is also possible that avid visitors have a more readily available mental list of recreational destinations than those who travel less frequently. In fact, one should also consider the possibility that those visitors most interested in wildlife-based outdoor recreation might have actually chosen a residence location close to a recreational site that offers that type of recreational activity.[5]

Visitors were asked about whether they had visited a series of alternative recreational sites in Atlantic Canada. The variable $TerraNova$, enters the final model with a negative sign. It makes sense that those coming to Gros Morne for the first time during the current trip from outside Newfoundland were more likely to take advantage of the trip to also visit Terra Nova National Park. More experienced and knowledgeable visitors were perhaps less likely to visit Terra Nova, suggesting that Gros Morne is the preferred choice among most people who have experienced both sites. *Highlands* has also a negative sign for mainly the same reason, but its effect is non-significant. Additionally, informal conversations with visitors during the sampling process revealed that it was very common for visitors from the US and Canadians from west of New Brunswick to drive through Cape Breton if they were visiting the Maritime Provinces for the first time. More frequent visitors of Gros Morne would then be more likely to drive or fly directly to Newfoundland.

6.1. Welfare Calculations

Manipulation of the results in Table 2 makes it possible to calculate consumer surplus measures for those visitors who went to Gros Morne for wildlife-based recreation. The final model accounts for the truncated and endogenous stratification of the data, so these welfare measures for a given population could be calculated, provided population values for the parameters in the demand equations were available (Englin & Shonkwiler 1995a). However, truncated individual models of recreation demand can be used to extrapolate welfare measures to non-visitors only under the assumption that these have the same demand functions

---

[5]This problem of endogeneity when referred to the travel cost to the site valued is, of course, well known and represents one of the most untractable shortcomings of the travel cost method (Parsons, 1991; Randall, 1994).

Table 3. Estimated welfare measures for visitors strongly interested in wildlife-based recreation at Gros Morne

	NBSTRAT	GNBSTRAT
$\widetilde{\beta}_{CTC}$	-0.4372	-0.2006
CS/persontrip[a]	$2,287	$4,985
CS/group for five years	$9,381	$21,753
CS/individual[b] for five years	$3,633	$8,425
Expected $persontrip$	4.1[c]	4.4[c]

(a) CS/persontrip = $1000(1/\beta_{CTC})$
(b) Based on an avegage partysize of 2.5821
(c) Based on $E(persontrip_i/x_i) = \lambda_i + 1 + \alpha_i \lambda_i$

as visitors (Hellerstein 1991). Since it is not entirely clear that this is the case, and it is unclear how the relevant population should be defined in this case, let alone how values for most of the demand parameters could be obtained for such a heterogeneous population, the calculations in this section refer only to users.[6] Another caveat is that only the estimated coefficients on the combination of $travelcost$ and $traveltimecost$ (that is CTC) are used to calculate welfare measures. This is because $expenses$ are mainly endogenous, a choice of the visitor. Although expenses could include some component of user fees, these are usually relatively small compared with the full cost of the visit. In any event, the welfare measures considered can be seen as a conservative lower bound for the full benefit derived by users.

From the results reported in Table 2 the consumer surplus per visit can be calculated as $-1/\beta_{CTC}$ (Creel & Loomis, 1990). If this expression is multiplied by predicted $persontrip$, we obtain the predicted CS per five year period for the typical visitor group in the sample results (Englin et al., 2003). This is the correct measure for policy analysis if it is assumed that the dominant source of error in the analysis is measurement error (Bockstael & Strand, 1987; Haab & McConnell, 2002, p. 162). The predicted mean $persontrip$ can be calculated by aggregating over all visitors and calculating the average count. If instead the error were expected to be mainly specification error, $-1/\beta_{TC}$ should be multiplied by the sample average.

As shown in Table 3 the value of consumer surplus per $persontrip$ under $NBSTRAT$ would be $2,287, accounting for the fact that all the cost variables were measured in thousands or dollars. The extrapolation to the consumer surplus per visiting group for the five years would accordingly be $9,381. When the overdispersion parameter α is allowed to vary across visitors (under $GNBSTRAT$), consumer surplus per $persontrip$ is instead $4,985. The consumer surplus per visiting group for the five years based on the predicted number of $persontrip$, rather than the observed number, can be calculated following the method in Englin and Shonkwiler (1995a), whereby $E(persontrip_i/x_i) = \lambda_i + 1 + \alpha_i \lambda_i =$

---

[6]See Shonkwiler and Shaw (1996) for a discussion of whether or not welfare measures calculated from a sample of users can be extrapolated to the general population. The manipulations that one would need to apply to estimated individual welfare measures to correct the on-site bias when calculating population-wide measures are available in Parsons (2003).

4.4. Marshallian surplus is $21,753$ per $partysize$ during the five previous years. Dividing by the sample mean of $partysize$ times five (years) we obtain a consumer surplus of $21,753/(2.5821 \cdot 5) = \$1,685$ per year per typical party member.

Note that the consumer surplus for the typical $partysize$ among those 540 visitors who stated instead that the influence of wildlife had an influence of 7 less in their decision to make the trip is only $2,819 per trip and therefore much less than the $4,985 equivalent found for the average party of those most interested in wildlife.

7. Conclusion

In this chapter, on-site survey data from Gros Morne National Park in Newfoundland were used to estimate truncated count data models of wildlife-based recreation demand. The preferred model, which accounts at the same time for overdispersion, zero-truncation and endogenous stratification in the distribution of the dependent variable yields a value of consumer surplus that confirms that the wildlife-based recreational amenities at the park are highly valued by visitors. For Gros Morne National Park, the results confirm that those visitors who are particularly interested in the recreational opportunities for encountering wildlife in the park derive substantially larger benefits from the park than other visitors for whom the availability of these opportunities did not have a major influence on their decision to take the trip. This is likely to be the case in similar recreational areas located far from urban centers and those that are home to rare wildlife species, or species that are rare for the latitude of the site. We also find that visits to the park are a normal good, which means that the demand for visits made for the purpose of wildlife-based recreation increases with visitors' income.

This type of result should be considered when making management decisions that might affect the abundance, or accessibility of wildlife species in recreational areas. In particular, park managers should take into consideration that highly visible wildlife species, such as moose, are usually the ones that provide the most benefits for visitors.

Acknowledgements

The author would like to thank Joe Amoako-Tuffour for his input during the overarching Gros Morne research project, Joe Hilbe (for help with the STATA codes), Jeff Anderson, Danny Major, Gareth Horne, Colleen Kennedy, Ken Kennedy, Dave Lough, John Gibbons, Bob Hicks, Paul Parsons, and the Parks Canada staff at Gros Morne. Courtney Casey, Jarret Hann, Perry Payne, Tracy Shears, and Thomas Khattar did an excellent job administering the survey, and Brian O'Shea provided invaluable research assistance. The survey effort was made possible by funding and/or logistic support provided by Parks Canada, ACOA, the Gros Morne Co-Operating Association, the Viking Trail Tourism Association, and SSHRC funds through a grant of the Centre of Regional Studies at St. Francis Xavier University.

References

Amoako-Tuffour, J., & Martínez-Espiñeira, R. (2008). *Leisure and the opportunity cost of travel time in recreation demand analysis.* (Mimeo St. Francis Xavier University. Available at http://ideas.repec.org/p/pra/mprapa/8573.html)

Bayless, D. S., Bergstrom, J., Messonnier, M., & Cordell, H. (1994). Assessing the demand for designated wildlife viewing sites. *Journal of Hospitality and Leisure Marketing*, **2**(3), 75-93.

Beal, D. (1995). A travel cost analysis of the value of Carnarvon Gorge National Park for recreational use. *Review of Marketing and Agricultural Economics*, **63**(2), 292-303.

Bell, F. W., & Leeworthy, V. R. (1990). Recreational demand by tourists for saltwater beach days. *Journal of Environmental Economics and Management*, **18**(3), 189-205.

Bhat, M. G. (2003). Application of non-market valuation to the Florida Keys Marine Reserve Management. *Journal of Environmental Management*, **67**(4), 315-325.

Bin, O., Landry, C. E., Ellis, C., & Vogelsong, H. (2005). Some consumer surplus estimates for North Carolina beaches. *Marine Resource Economics*, **20**(2), 145-161.

Bockstael, N. E., & Ivar E. Strand, J. (1987). The effect of common sources of regression error on benefit estimates. *Land Economics*, **63**(1), 11-20.

Bowker, J. M., English, D. B. K., & Donovan, J. A. (1996). Toward a value for guided rafting on southern rivers. *Journal of Agricultural and Applied Economics*, **28**(2), 423-432.

Braden, J. B., & Kolstad, C. (1991). *Measuring the demand for environmental quality.* Amsterdam: Elsevier.

Cameron, A. C., & Trivedi, P. K. (1990). Regression-based tests for overdispersion in the poisson model. *Journal of Econometrics*, **46**(3), 347-364.

Cameron, A. C., & Trivedi, P. K. (2001). Essentials of count data regression. In B. H. Baltagi (Ed.), *A companion to theoretical econometrics* (p. 331-348). Oxford, U.K.: Blackwell.

Cameron, C., & Trivedi, P. K. (1998). *Regression analysis of count data.* Cambridge: Cambridge University Press.

Cesario, F. (1976). Value of time in recreation benefit studies. *Land Economics*, **52**, 32-41.

Cooper, J., & Loomis, J. (1991). Economic value of wildlife resources in the San Joaquin Valley: Hunting and viewing values. In A. Dinar & D. Zilberman (Eds.), *The economic and management of water and drainage in agriculture* (p. 447-463). Boston, MA: Kluwer Academic Publishers.

Creel, M., & Loomis, J. B. (1990). Theoretical and empirical advantages of truncated count data estimators for analysis of deer hunting in California. *American Journal of Agricultural Economics*, **72**, 434-441.

Curtis, J. A. (2002). Estimating the demand for salmon angling in Ireland. *The Economic and Social Review*, **33**(3), 319-332.

Dobbs, I. M. (1993). Adjusting for sample selection bias in the individual travel cost method. *Journal of Agricultural Economics*, **44**, 335-342.

D.W. Knight Associates. (2005). *Gros Morne National Park visitor assessment 2004* (Tech. Rep.). Parks Canada.

Earnhart, D. (2004). Time is money: Improved valuation of time and transportation costs. *Environmental and Resurce Economics*, **29**(2), 159-190.

Englin, J., Holmes, T. P., & Sills, E. O. (2003). Estimating forest recreation demand using count data models. In E. O. Sills (Ed.), *Forests in a market economy* (p. 341-359). Dordrecht, The Netherlands: Kluwer Academic Publishers.

Englin, J., & Shonkwiler, J. (1995a). Estimating social welfare using count data models: An application under conditions of endogenous stratification and truncation. *Review of Economics and Statistics*, **77**, 104-112.

Englin, J., & Shonkwiler, J. S. (1995b). Modeling recreation demand in the presence of unobservable travel costs: Toward a travel price model. *Journal of Environmental Economics and Management*, **29**(3), 368-377.

Feather, P., & Shaw, W. D. (1999). Estimating the cost of leisure time for recreation demand models. *Journal of Environmental Economics and Management*, **38**(1), 49-65.

Fix, P., & Loomis, J. (1997). The economic benefits of mountain biking at one of its meccas: An application of the travel cost method to mountain biking in Moab, Utah. *Journal of Leisure Research*, **29**(3), 342-352.

Fix, P., & Loomis, J. (1998). Comparing the economic value of mountain biking estimated using revealed and stated preference. *Journal of Environmental Planning and Management*, **41**(2), 227- 236.

Freeman III, A. M. (1993). *The measurement of environmental and resource values: Theory and methods*. Washington D.C.: Resources for the Future.

Garrod, G. D., & Willis, K. G. (1999). *Economic valuation of the environment*. Cheltenham: Edward Elgar.

Grogger, J. T., & Carson, R. T. (1991). Models for truncated counts. *Journal of Applied Econometrics*, **6**(3), 225-238.

Gürlük, S., & Rehber, E. (2008). A travel cost study to estimate recreational value for a bird refuge at Lake Manyas, Turkey. *Journal of Environmental Management*, **88**(4), 1350-1360.

Gurmu, S., & Trivedi, P. (1996). Excess zeros in count models for recreational trips. *Journal of Business and Economic Statistics*, **14**, 469-477.

Haab, T., & McConnell, K. (2002). *Valuing environmental and natural resources: Econometrics of non-market valuation*. Cheltenham, UK: Edward Elgar.

Hagerty, D., & Moeltner, K. (2005). Specification of driving costs in models of recreation demand. *Land Economics*, **81**(1), 127-143.

Heberling, M. T., & Templeton, J. J. (2008). Estimating the economic value of national parks with count data models using on-site, secondary data: The case of the Great Sand Dunes National Park and Preserve. *Environmental Management, Published online: 28 May 2008*, forthcoming.

Hellerstein, D., & Mendelsohn, R. (1993). A theoretical foundation for count data models. *American Journal of Agricultural Economics*, **75**(3), 604-611.

Hellerstein, D. M. (1991). Using count data models in travel cost analysis with aggregate data. *American Journal of Agricultural Economics*, **73**, 860-866.

Hesseln, H., Loomis, J. B., González-Cabán, A., & Alexander, S. (2003). Wildfire effects on hiking and biking demand in New Mexico: A travel cost study. *Journal of Environmental Management,*, **69**(4), 359-368.

Hilbe, J. (2007). *Negative binomial regression.* Cambridge, UK: Cambridge University Press. (In Press)

Hilbe, J. M. (2005). *GNBSTRAT: Stata module to estimate generalized negative binomial with endogenous stratification.* Statistical Software Components, Boston College Department of Economics. Statistical Software Components, Boston College Department of Economics. (Available online at http://ideas.repec.org/c/boc/bocode/s456413.html)

Hilbe, J. M., & Martínez-Espiñeira, R. (2005). *NBSTRAT: Stata module to estimate negative binomial with endogenous stratification.* Statistical Software Components, Boston College Department of Economics. (Available online at http://econpapers.repec.org/software/bocbocode/s456414.htm)

Krutilla, J. V. (1967). Conservation reconsidered. *The American Economic Review,* **57**(4), 777-786.

Leeworthy, V. R., & Bowker, J. M. (1997). *Nonmarket economic user values of the Florida Keys/Key West* (Tech. Rep.). Silver Spring, MD: National Oceanic and Atmospheric Administration, Strategic Environmental Assessments Division.

Liston-Heyes, C., & Heyes, A. (1999). Recreational benefits from the Dartmoor National Park. *Journal of Environmental Management,* **55**(2), 69-80.

Locke, W., & Lintner, A. M. (1997). *Benefits of protected areas: The Gros Morne National Park case study* (Tech. Rep.). Parks Canada.

Loomis, J. (2003). Travel cost demand model based river recreation benefit estimates with on-site and household surveys: Comparative results and a correction procedure. *Water Resources Research,* **39**(4), 1105.

Mäler, K. (1974). *Environmental economics: A theoretical inquiry. resources for the future.* Baltimore, MD: Johns Hopkins University Press.

Marsinko, A., Zawacki, W. T., & Bowker, J. M. (2002). Use of travel cost models in planning: A case study. *Tourism Analysis,* **6**, 203-211.

Martínez-Espiñeira, R., & Amoako-Tuffour, J. (2008a). *Multi-destination and multi-purpose trip effects in the analysis of the demand for trips to a remote recreational site.* (42nd Annual Meetings of the Canadian Economic Association, University of British Columbia, Vancouver, June)

Martínez-Espiñeira, R., & Amoako-Tuffour, J. (2008b). Recreation demand analysis under truncation, overdispersion, and endogenous stratification: An application to Gros Morne National Park [Econometrics]. *Journal of Environmental Management,* **88**(4), 1320-1332.

Martínez-Espiñeira, R., Amoako-Tuffour, J., & Hilbe, J. M. (2006). Travel cost demand model based river recreation benefit estimates with on-site and household surveys: Comparative results and a correction procedure: Reevaluation. *Water Resources Research,* **42**(W10418). (doi:10.1029/2005WR004798)

Martínez-Espiñeira, R., & Hilbe, J. M. (2008). Effect on recreation benefit estimates from correcting for on-site sampling biases and heterogeneous trip overdispersion in count data recreation demand models. *Journal of Modern Applied Statistical Methods,* forthcoming.

Martínez-Espiñeira, R., Loomis, J. B., Amoako-Tuffour, J., & M.Hilbe, J. (2008). Comparing recreation benefits from on-site versus household surveys in count data travel cost demand models with overdispersion. *Tourism Economics*, **14**(3), 567-576.

McKean, J. R., Johnson, D., & Taylor, R. G. (2003). Measuring demand for flat water recreation using a Two-Stage/Disequilibrium travel cost model with adjustment for overdispersion and self-selection. *Water Resources Research*, **39**(4), 1107.

Moeltner, K., & Shonkwiler, J. S. (2005). Correcting for size biased sampling in random utility models. *American Journal of Agricultural Economics*, **87**(2), 327-339.

Ovaskainen, V., Mikkola, J., & Pouta, E. (2001). Estimating recreation demand with on-site data: An application of truncated and endogenously stratified count data models. *Journal of Forest Economics*, **7**(2), 125-144.

Parks Canada. (2004a). *Gros Morne National Park traffic study and attendance formula 2004* (Tech. Rep.). Management Planning and Social Services Section, Atlantic Service Centre.

Parks Canada. (2004b). *Gros Morne National Park visitor study 2004* (Tech. Rep.). Management Planning and Social Services Section, Atlantic Service Centre.

Parsons, G. (1991). A note on choice of residential location in travel cost demand models. *Land Economics*, **67**(3), 360-364.

Parsons, G. R. (2003). The travel cost model. In P. A. Champ, K. J. Boyle, & T. C. Brown (Eds.), *A primer on nonmarket valuation* (chap. 9). London: Kluwer Academic Publishing.

Randall, A. (1994). A difficulty with the travel cost method. *Land Economics*, **70**(1), 88-96.

Sarker, R., & Surry, Y. (1998). Economic value of big game hunting: The case of moose hunting in Ontario. *Journal of Forest Economics*, **4**(1), 29-60.

Shaw, D. (1988). On-site sample regression: Problems of non-negative integers, truncation, and endogenous stratification. *Journal of Econometrics*, **37**, 211-223.

Shonkwiler, J., & Shaw, W. (1996). Hurdle count-data models in recreation demand analysis. *Journal of Agricultural and Resource Economics*, **21**(2), 210-219.

Shrestha, K. R., Seidl, A. F., & Moraes, A. S. (2002). Value of recreational fishing in the Brazilian Pantanal: A travel cost analysis using count data models. *Ecological Economics*, **42**(1-2), 289-299.

Siderelis, C., & Moore, R. L. (1995). Outdoor recreation net benefits of rail-trails. *Journal of Leisure Research*, **27**(4), 344-359.

Smith, V. K., Desvousges, W., & McGivney, M. (1983). The opportunity cost of travel time in recreation demand models. *Land Economics*, **59**(3), 259-278.

Sohngen, B., Lichtkoppler, F., & Bielen, M. (2000). *The value of day trips to Lake Erie beaches* (Tech. Rep. No. TB-039). Columbus OH: Ohio Sea Grant Extension.

Ward, F. A., & Loomis, J. B. (1986). The travel cost demand model as an environmental policy assessment tool: A review of literature. *Western Journal of Agricultural Economics*, **11**(2), 164-178.

Yen, S. T., & Adamowicz, W. L. (1993). Statistical properties of welfare measures from count-data models of recreation demand. *Review of Agricultural Economics*, **15**, 203-215.

Zawacki, W. T. A. M., & Bowker, J. M. (2000). A travel cost analysis of nonconsumptive wildlife-associated recreation in the United States. *Forest Science*, **46**(4), 496-506.

INDEX

A

abiotic, 91, 94
Abiotic, 91
ACC, 151, 165
access, xii, 19, 27, 29, 30, 120, 121, 124, 144, 181, 182
accessibility, 30, 195
accommodation, 120, 121, 126, 127, 128, 129, 182
accounting, 60, 106, 188, 194
acetate, 167, 168
achievement, 121, 122, 129, 130
acid, 99, 148, 150, 162, 164
acidophilic, 159
acidotolerant, 159
Acinetobacter, 153
Actinobacteria, 152, 153, 154, 156, 168
actinomycosis, 113
adaptation, 17, 159
adjustment, 199
administration, 13, 19, 32, 44, 118, 127, 129
administrative, viii, 26, 43, 46, 47, 50, 51, 52, 54, 56, 126, 127
administrators, 30
adult, 98, 114
adults, 98
aerobic, 159, 167
aesthetics, 120
Africa, 14, 83, 84, 118, 125, 130, 149
AGC, 151, 165
age, 36, 49, 70, 184, 190
agent, 174
agents, 169
aggression, 139
aging, 69
agricultural, 62, 135
agriculture, 17, 28, 123, 124, 129, 196
air, 92, 94
airports, 188, 189
Alaska, 90
algorithm, 151, 165
alkaline, 124, 126

alkaloids, 92
Alps, 13, 14, 90, 92, 106
alternative, viii, 30, 33, 43, 44, 45, 48, 50, 51, 56, 128, 179, 185, 189, 193
alternatives, 26, 33, 44, 80, 81, 144
alters, 31
alveoli, 95
Amazon, 174
Amazonian, 12, 19
American culture, 26
ammonium, 167
amphibia, 23, 178
amphibians, 16, 68
Amsterdam, 196
anaerobe, 168
anaerobic, 159, 167
animal diseases, 176
animals, vii, ix, 85, 86, 89, 90, 92, 93, 94, 95, 97, 100, 101, 106, 109, 111, 114, 115, 120, 122, 125, 128, 139, 145, 147, 148, 162, 163, 167, 169, 176, 178
Animals, 111
Antarctic, 170
anthropic, 93, 94, 97, 106, 107, 108
anthropic factors, 106
anthropogenic, 17, 109, 132, 133, 145
application, 44, 57, 58, 59, 64, 68, 71, 72, 184, 186, 190, 197, 198, 199
aquatic systems, 22, 171
Archaea, 148, 158, 167, 168
argument, 14
Arizona, 28, 29, 35, 40, 41
articulation, 32
Asia, 90, 174
assessment, 18, 32, 33, 59, 67, 72, 76, 82, 83, 109, 196, 199
assets, xii, 69, 129, 139, 181
assumptions, 182
asymptotically, 184
Atlantic, xi, 163, 173, 174, 175, 189, 193, 199
atmosphere, xi, 167, 173
attachment, 20
attitudes, ix, 61, 62, 65, 70, 83, 84, 180
Australia, 14, 58, 165

Austria, 101, 103
authority, ix, 61, 62, 63, 65, 66, 67, 71, 72, 76, 82
autopsy, 97
availability, xi, 161, 195
awareness, 23, 144

B

Bacillus, 152, 153
bacteria, 106, 150, 151, 152, 154, 155, 162, 167, 171
bacterial, 106, 148, 150, 154, 155, 156, 162, 167
bacterium, 106, 159
Balkans, 90
barrier, 70
barriers, 15, 49
beaches, 31, 34, 39, 196, 199
beetles, 22
behavior, 63, 83, 122, 174, 176
behaviours, 175
beliefs, x, 20, 131, 132, 133, 140, 145
benefits, 27, 29, 30, 39, 60, 67, 79, 124, 144, 174, 181, 195, 197, 198, 199
benign, 174
Bhutan, x, 131, 134
bias, 28, 162, 189, 194, 196
big game hunting, 199
binomial distribution, 184, 186, 187
bioavailability, 171
biodegradation, 169
biodiversity, vii, viii, x, xi, xii, 1, 3, 14, 16, 18, 19, 20, 21, 22, 25, 27, 38, 39, 45, 49, 56, 61, 62, 68, 83, 84, 97, 106, 117, 121, 122, 123, 126, 127, 128, 129, 130, 132, 133, 142, 144, 146, 147, 149, 170, 173, 174, 175, 178, 179, 181
biogeography, 157, 162
biological processes, 38
biomass, 126, 167, 175
biopsy, 98
biosphere, 59, 87, 171
biotic, 2, 16, 21, 39, 93
biotic factor, 93
birds, 14, 16, 20, 22, 68
birth, 97
Black Sea, 46
blindness, 106
blocks, 124
blood, 100
blood stream, 100
boats, 30
body weight, 91
boiling, 148
borderline, 85
Boston, 196, 198
bounds, 28, 67, 79
Bradyrhizobium, 153
Brazil, xi, 58, 173, 174, 175, 176, 177, 178, 179, 180
Brazilian, xi, xii, 173, 174, 175, 178, 179, 199
breeding, 108

Britain, 23
British Columbia, 198
bronchioles, 95
Buddhism, 132, 139, 140
Buddhist, x, 132, 133, 135, 140
buffer, 13, 14, 15, 16, 17, 46, 47, 49, 84, 87
Bureau of Land Management, 26
Bureau of Reclamation, 27, 28, 31, 32, 33, 34, 35, 39, 40
Burkholderia, 153
burn, 175
burning, 135, 174, 178, 179

C

campaigns, 120
Canada, 181, 183, 184, 188, 193, 195, 196, 198, 199
candidates, 17
capacity, 32, 57, 71, 74, 75, 77, 119, 121, 144, 169, 175
carbon, 167, 168, 174
carbon dioxide, 167
Carpathian, 87, 88, 89, 91, 93, 101, 115
case law, 32
case study, vii, xi, 1, 16, 27, 58, 161, 198
catalyst, 70
cattle, 114
Caucasus, 90, 92, 113, 174
cave, 98
cell, 150
Census, 60, 134
Central Asia, 90, 174
Central Europe, 88
certificate, 45, 49
certification, 58
CGT, 151, 164
CH_4, 168
children, 26
China, x, 62, 84, 90, 131, 134
ciliate, 142
Ciliates, 168
circulation, 115
citizens, 135
classical, 157
classification, 150, 162
climate change, 18
climate warming, 2, 17
climatic factors, 100
cloning, 151, 165
CO_2, 168
Coccidia, 106
codes, 195
cognitive, 82
coleoptera, 19, 22
colonization, 149
Colorado, vii, viii, 25, 27, 28, 29, 30, 31, 32, 33, 34, 35, 36, 37, 38, 39, 40, 41, 42
Columbia, 83

commodity, 26
communication, 14, 16, 20, 93, 120, 121
communities, ix, x, xi, 15, 19, 21, 35, 38, 41, 61, 62, 63, 65, 69, 70, 84, 87, 124, 131, 132, 133, 139, 144, 145, 147, 148, 154, 156, 157, 158, 159, 161, 162, 163, 166, 167, 169, 170, 171
community, x, xi, 39, 62, 63, 64, 65, 70, 71, 72, 83, 84, 124, 130, 132, 140, 142, 143, 144, 147, 153, 154, 157, 159, 166, 167, 170, 171
compensation, 25, 123, 130
competition, 36, 126
complementarity, 183
complexity, 34, 162, 166, 170
components, 39, 151, 154, 157, 165, 167
composition, 87, 88, 100, 125, 126, 135, 140, 156, 184, 190
compounds, 167
computation, 74, 82
computer technology, 45
computing, 191
concentrates, 118
concentration, 118, 184
conception, 2, 22
concrete, 28
conditional mean, 185, 186
configuration, 91
conflict, vii, 25, 27, 33, 35, 38, 39, 129
confusion, 32
Congress, 26, 32, 39, 115
conifer, 134
coniferous, 90, 135
consciousness, 28, 144
consensus, 86, 142
conservation, vii, viii, ix, x, xi, xii, 1, 2, 3, 13, 14, 15, 16, 17, 18, 19, 20, 21, 22, 23, 25, 43, 45, 46, 52, 54, 55, 56, 57, 61, 62, 63, 64, 65, 66, 67, 68, 70, 71, 72, 76, 77, 78, 79, 80, 81, 82, 83, 84, 86, 87, 93, 107, 108, 113, 115, 117, 118, 119, 120, 121, 122, 123, 124, 126, 127, 128, 129, 130, 132, 135, 140, 142, 143, 144, 145, 146, 148, 149, 157, 161, 167, 170, 173, 175, 178, 179
constraints, ix, 61, 62
construction, 28, 35, 36, 37, 38, 39, 76, 145, 188
consulting, 135
consumer surplus, xii, 181, 182, 183, 191, 193, 194, 195, 196
consumers, 48
consumption, 168, 183
contamination, 97, 154
contractions, 113
contracts, 31
control, 18, 28, 36, 37, 44, 62, 69, 94, 127, 135, 140, 144, 169, 178, 179
cooling, 92
coral, 15, 16
coral reefs, 15
correlation, 48, 66, 78
correlations, 47, 50, 65, 66, 72, 78, 82
Corynebacterium, 152, 154
costs, 67, 79, 82, 123, 181, 182, 188, 189, 197
coupling, 13
courts, 31, 32, 35
covering, x, 87, 92, 117, 131, 134, 163, 165
credit, 58
credit card, 58
crime, 120
critical habitat, 17, 35
criticism, 35
crossbreeding, 86
crust, 158
crustaceans, 68
crystalline, 87
CSR, 111
cultivation, 132, 135
cultural heritage, x, 13, 68, 72, 132, 144, 145
cultural practices, 139
cultural values, 57, 59
culture, x, 69, 132, 144, 151, 162, 167
customers, 28
Cyanobacteria, 168
cycles, xi, 147, 151, 165, 167, 175
cycling, xi, 161
cyclones, 132
Czech Republic, 90, 101

D

danger, 30, 38, 92, 121, 180
Darjeeling, 132, 140, 146
data collection, 182
data distribution, 184
database, 151
dating, 37, 89, 90
death, 26, 106, 128
debt, 17, 23
decay, 155, 156
deciduous, 140
decision makers, 18
decision making, 44, 45, 144
decisions, 26, 35, 55, 195
deep-sea, 159
definition, 46, 187, 189
degradation, 16, 17, 94, 118, 126, 133, 145, 169, 171, 172, 174
degrading, 123
demand, xii, 29, 181, 182, 183, 184, 187, 188, 189, 192, 193, 194, 195, 196, 197, 198, 199
demand curve, 183, 192
democracy, 55
denaturation, 158
denaturing gradient gel electrophoresis, 158, 171
denitrification, 168
density, 91, 99, 106, 109, 118, 119, 120, 124, 125, 126, 129, 183, 184, 185, 186, 187
Department of the Interior, 42
dependent variable, xii, 48, 51, 181, 182, 188, 190, 195

depression, ix, 85
desert, 28
desiccation, 163
desire, 18, 55, 124
destruction, 16, 17, 23, 36, 126, 129, 132, 174
detection, xi, 147, 150, 151, 155, 162, 164
developed countries, 45
developing countries, 83, 84
Diamond, 2, 21, 36, 37, 100
dichotomy, 19
diet, 110, 111
differential diagnosis, 95
differentiation, 87
digestion, 92
direct observation, 148
disappointment, 33
discourse, 66
diseases, 94, 125
dispersion, xi, 17, 147, 148, 156, 157, 185
displacement, 126, 158, 171, 175
disposition, 63
distribution, xii, 17, 21, 36, 38, 87, 91, 99, 154, 162, 171, 178, 181, 182, 184, 185, 186, 187, 190, 191, 195
diversification, 113
diversity, vii, x, xi, xii, 2, 7, 15, 16, 17, 20, 22, 23, 85, 86, 87, 99, 100, 109, 118, 122, 124, 125, 126, 127, 129, 131, 133, 140, 145, 146, 147, 148, 149, 157, 158, 161, 162, 163, 167, 169, 170, 171, 173, 175
division, 124, 129
DNA, 150, 151, 152, 153, 154, 164, 165
dominance, 129
doors, 27
draft, 44
drainage, 196
droughts, 38
drying, 99
duality, 19
duration, 184, 189, 193
dust, 130
duties, 32

E

earth, 39, 94, 179
ecological, viii, ix, 2, 17, 18, 23, 27, 30, 31, 32, 33, 35, 38, 39, 40, 43, 44, 45, 57, 68, 69, 85, 86, 87, 106, 107, 112, 126, 130, 132, 133, 174
Ecological Economics, 83, 84
ecological restoration, 32, 33
ecology, 69, 95, 107, 109, 113, 143, 179
econometric analysis, 182
econometrics, 196
economic activity, 29
economic development, 29, 83
economic growth, 27
economic incentives, 84, 146

economics, 198
ecosystem, x, xi, 2, 13, 17, 22, 27, 30, 32, 34, 58, 60, 86, 117, 123, 124, 143, 148, 150, 157, 161, 169, 170, 173, 174, 175, 178, 179
ecosystem restoration, 32, 34
ecosystems, vii, xi, 25, 26, 27, 28, 30, 93, 106, 115, 130, 143, 147, 157, 161, 163, 167, 169, 170, 171, 174, 178, 179
Education, 50, 72, 77, 79
educational attainment, 190
effluents, 127
elasticity, 57
elderly, 135
electricity, 28, 31, 35, 39
electrophoresis, 154, 158, 171
elephant, x, 117, 118, 124, 125, 126, 129, 130
employment, 62, 144
empowered, 63
Endangered Species Act, 21, 31
endogeneity, 193
energy, 15, 17, 33, 167
entrepreneurs, 144
entropy, 63
environment, vii, xi, 15, 27, 28, 31, 32, 58, 69, 83, 90, 91, 92, 97, 100, 101, 109, 121, 124, 126, 129, 143, 147, 148, 150, 156, 157, 161, 162, 163, 165, 166, 167, 169, 170, 175, 197
environmental change, 163, 167
environmental conditions, 17, 62
environmental degradation, 169
environmental factors, 140, 157
environmental impact, 33
environmental policy, 199
environmental protection, 163
environmental threats, 17
environmentalists, 32, 33, 34
equilibrium, xi, 147, 161, 163, 169
equipment, 49, 193
erosion, 31, 34, 62
Escherichia coli, 155, 156
EST, 133
estimating, 189
estimator, 185
estimators, 196
ethnic groups, 133
ethnicity, 144
Europe, 45, 46, 56, 85, 89, 90, 95, 112, 115, 174
evaporation, 88
evapotranspiration, 163
evening, 68, 94
evolution, 41, 159, 169, 171, 172
ewe, 115
exaggeration, 20
exclusion, 16, 174
execution, 144
exercise, 135
expansions, 113
expenditures, 29, 60
exploitation, ix, 32, 85, 86, 122

exposure, xi, 125, 147, 155, 176
external environment, 48
extinction, ix, 21, 22, 23, 85, 86, 93, 126, 130, 148, 169, 172
extraction, 135, 150, 164
extrapolation, 194
extrusion, 95, 96

F

faecal, 97, 98
failure, 35
faith, 135, 139, 140, 145
family, 19, 90, 95, 140
FAO, 46, 49, 58, 179
fatalities, 30
fauna, 14, 45, 49, 56, 86, 87, 88, 95, 99, 101, 108, 110, 111, 112, 113, 170, 178, 179, 180
fear, 135
February, 59
feces, 113
federal courts, 35
fee, viii, 43, 47, 49, 50, 51, 52, 54, 55, 56, 57, 182
feedback, 33
feeding, 121, 125
fees, 181, 194
feet, 28, 34, 36
fencing, 124
fibrosis, 100
film, 118, 165, 166, 167
finance, 49
fingerprinting, 151, 165
fire, xi, xii, 60, 173, 174, 175, 176, 178, 179, 180
fire event, xi, 173, 175, 178, 179
fire suppression, 174, 179
fire-adapted, 174
fires, xi, 132, 173, 174, 175, 179
fish, vii, viii, 25, 30, 31, 34, 35, 38, 39
Fish and Wildlife Service (FWS), 31, 32, 33, 34, 35
fishing, 29, 135, 144, 190, 199
fixation, 168
flexibility, 17
flight, 188
flood, 28, 31, 33, 34, 35, 36, 37, 38, 175
flooding, 28, 39, 41
flora, viii, 14, 25, 36, 45, 49, 56, 85, 87, 88, 170, 178
flora and fauna, 49, 170
flow, 30, 31, 33, 35, 36, 37
fluctuations, 30, 31, 37, 38, 87
fluvial, 175
FMA, 180
focusing, 2, 18, 36, 144
food, 90, 91, 92, 99, 176
forest fire, 132
forest fires, 132
forest management, x, 58, 131, 132
Forestry, 43, 45, 57, 58, 59, 60, 62, 68, 83, 84, 131, 174, 179
forests, x, 15, 17, 22, 45, 49, 56, 86, 88, 90, 99, 107, 131, 132, 133, 134, 135, 140, 145
fossil, 90
Fox, 21, 23
fragmentation, 17, 22, 128, 172
France, 1, 6, 13, 14, 15, 16, 21, 22, 23, 101, 103, 108, 109, 110, 115
freedom, 21
freezing, 92
freshwater, xi, 88, 161, 165, 166, 167
fuel, 145, 174, 175, 178, 179
funding, 195
funds, 146, 195
fungi, 148
fuzzy logic, 63, 65, 67, 76, 79, 82
fuzzy sets, 65

G

garbage, 127
gas, 107
gastric, 113
gastrointestinal, 94, 100, 101, 103, 104, 106, 108
gel, 158, 171
GenBank, 153, 154, 165
gender, 70
gene, 97, 111, 150, 153, 154
generation, 14, 28, 32, 33, 35, 36, 39, 118, 129
genes, 107, 151, 164
genetic diversity, 17
genetics, 115
geochemical, xi, 161
geology, 140
geothermal, 158
geothermal fluids, 158
Germany, 90, 150, 151, 164, 165
GFI, 180
glaciation, 87, 89, 90
glaciers, 87
global climate change, 94, 97
Global Warming, 18, 41
goals, vii, viii, 1, 17, 19, 32, 38, 61, 62, 126, 130, 170
goods and services, 45, 182
government, ix, x, 29, 61, 69, 72, 117, 118, 124, 126, 130, 132, 135, 144
GPS, 135
granites, 87
grants, 124
grass, 92, 124
grassland, 124
grasslands, 17, 23, 124, 125, 126, 174, 175
grazing, 91, 92, 135
Great Depression, 28
greed, 16
groups, viii, x, 38, 43, 44, 45, 49, 52, 53, 54, 56, 57, 62, 70, 89, 90, 94, 100, 131, 133, 145, 148, 150, 152, 155, 156, 162, 167, 168, 175

growth, 27, 36, 37, 38, 62, 93, 148, 152, 156, 162, 167
growth rate, 37
growth temperature, 148
guidelines, vii, 1, 2, 63, 64, 65, 68, 70, 71, 82, 179

H

habitat, viii, 15, 16, 17, 22, 25, 27, 28, 30, 31, 32, 33, 34, 35, 36, 39, 91, 94, 124, 125, 126, 127, 128, 175, 179
handling, 154
harassment, 121, 128
harm, 13, 19, 31, 32, 33, 34, 38, 92
harmony, 13, 19
harvest, 22
harvesting, 18
Hawaii, 84
healing, 92, 93
health, 36, 37, 100, 101, 106, 120, 174, 179
heart, 128
height, 36
herbivores, 126
herbs, 140, 141
heterogeneity, 86, 126, 185
heterogeneous, 47, 62, 194, 198
heteroskedasticity, 184, 185
heterotrophic, 168
hibernation, 99
high risk, 157
high temperature, 147, 148, 149, 150, 151, 152, 155, 156
holistic, 17
Holocene, 90
homogenous, 100
host, 99, 109
hot spots, 2, 7, 15, 46, 49, 59, 156
hot spring, xi, 147, 148, 150
hotels, 126, 127
House, 64, 113, 115
household, 140, 198
households, 70
human, viii, ix, x, xi, 13, 14, 15, 17, 18, 19, 23, 26, 27, 28, 61, 65, 83, 93, 94, 107, 110, 117, 122, 123, 124, 127, 128, 129, 133, 134, 145, 148, 157, 169, 173, 174, 175
human actions, 175
human activity, 23, 27
human behavior, 83
human development, 13, 14
humans, 3, 15, 143, 169, 179
humidity, 98, 99, 100
humility, 26
Hungary, 93
hunting, ix, 13, 58, 59, 85, 86, 93, 94, 95, 115, 126, 135, 144, 176, 196, 199
hydrocarbon, 172
hydrogen, 167
hydrothermal, 159
hypothermia, 30
hypothesis, 23, 186

I

IAP, 180
ICDPs, 62, 83
ice, 150, 164
identification, 15, 17, 33, 82, 153, 154, 167
identity, 20, 62
images, 16
imagination, 26
immigrants, 157
immigration, 123, 148, 156, 157
implementation, ix, 34, 61, 64, 65, 67, 79, 82
in situ, 150
incentive, 62, 144
incentives, 62, 83, 145
income, viii, 27, 43, 45, 47, 49, 50, 51, 52, 54, 55, 56, 57, 144, 184, 188, 189, 190, 191, 192, 193, 195
incomes, 49, 55, 57, 144
incompatibility, 32
incubation, 151, 165
independent variable, 48, 188
India, x, 131, 132, 133, 134, 135, 139, 141, 145, 146
Indian, x, 132, 146
indication, 191
indicators, 18, 22, 71, 182
indices, 122
indigenous, ix, 61, 62, 63, 83, 84, 132, 133, 140, 145
indigenous peoples, 63, 84
individual perception, 20
Indonesia, 84
industry, viii, 25, 29, 30, 39, 58, 59, 61, 62, 63, 64, 118, 124, 126
infection, 95, 97, 98, 99, 100, 106, 109, 113
infections, 97, 99, 109, 113, 115
infectious, 94
infectious disease, 94
infectious diseases, 94
infestations, 110
Information System, 2
infrastructure, x, 117, 120, 126, 127, 129
ingest, 92
inhibition, 169
injuries, 94
innovation, 57
inorganic, 167
insects, 17
insecurity, 121
instability, 158
institutions, 49, 135, 142
instruments, 20
insurance, 17
intangible, 144
integration, 20

integrity, xi, 27, 30, 147
intensity, 174, 179, 185
intentions, 70
interactions, 2, 16, 135, 162
interdependence, 79
interdisciplinary, 129
interest groups, 38, 44, 45, 49, 52, 53, 54, 56, 57
interference, 93, 97
interpretation, 70
interrelations, 114
interstate, 27
intervention, 32, 82, 148, 157
interview, 45, 47, 48, 51, 70
interviews, 70, 71, 82
intrinsic, 21, 183, 184
intrinsic value, 183
investment, 71, 72, 79
ions, 17
Ireland, 196
iron, 165, 167, 171
irrigation, 62, 129
island, 149
isolation, x, 20, 99, 131, 133, 148, 183
Italy, 90, 101, 103, 106
IUCN, ix, 13, 22, 23, 85, 86, 107, 108, 130

J

Japan, 59, 62
job creation, 29
jobs, 19, 28, 29
journalists, 13
judge, 14, 76
judgment, 62, 73, 74, 75, 76, 79
Jung, 136, 137, 138
jurisdiction, 38, 123
jurisdictions, 30

K

kelp, 15
Kenya, ix, 117, 118, 119, 120, 121, 122, 123, 126, 129, 130
killing, 118, 124
Kobe, 59
Korea, 60, 62, 90

L

lakes, 95
land, x, 2, 15, 16, 18, 20, 27, 31, 39, 48, 86, 117, 122, 123, 124, 125, 128, 129, 131, 134, 135, 144, 175, 176, 183
land use, x, 17, 18, 27, 117, 123, 124, 125, 129, 134
Land Use Policy, 129

landscapes, vii, xi, 4, 6, 14, 15, 17, 19, 21, 26, 59, 147, 157, 163, 169
language, 70
language barrier, 70
Laos, 59
large-scale, 16, 158
larvae, 96, 97, 98, 99, 100, 109, 113
larval, 97, 98, 101
law, 26, 32, 33, 86, 93
laws, 28, 31, 32
lead, vii, xi, 16, 17, 19, 147, 161, 167, 169
leakage, 29, 144
legal protection, 15
legislation, 28
leisure, 197
leisure time, 197
lesions, 175
liberal, ix, 85, 86
life cycle, 99
lifestyle, 145
lifetime, 93, 100
likelihood, 70, 184, 185, 186, 187, 191
Likert scale, 47
limestones, 87, 88
Lincoln, 41, 58
linear, 185
linear regression, 185
linguistic, 65, 66, 67, 68, 70, 72, 73, 74, 76, 79, 80, 82
linguistics, 76, 78
liquidate, 86
livestock, x, 117, 118, 125, 126, 129
living conditions, 99, 100, 168
living standard, 145
living standards, 145
lobbying, 13
local community, 144
local government, 72, 144
location, 95, 120, 127, 128, 135, 149, 154, 166, 184, 188, 192, 193, 199
logistics, 40
London, 58, 108, 109, 171, 199
longevity, 28
longitudinal study, 83
long-term, xi, 18, 36, 44, 142, 161
Los Angeles, 42
losses, xi, 17, 93, 95, 101, 107, 110, 123, 173, 174, 178
Louisiana, 58
love, 26
lover, 97, 99
low temperatures, 99
lung, 94, 95, 96, 97, 98, 99, 100, 109, 113, 114
lungs, 95, 98, 100, 109, 114
lupus, 6, 13, 14, 111, 114, 115
lysis, 162

M

machinery, 126
macroorganisms, 148, 167
magma, 158
mainstream, 39
maintenance, vii, xi, 115, 126, 147, 148, 161, 163, 167
males, 95
mammal, 120, 125, 126, 175
mammals, 14, 68, 124, 126, 130, 175, 178, 183
management, vii, viii, ix, x, xii, 1, 2, 18, 20, 21, 22, 25, 26, 27, 28, 30, 31, 32, 33, 34, 35, 38, 39, 43, 44, 45, 46, 47, 48, 49, 50, 51, 54, 55, 56, 57, 58, 59, 60, 61, 62, 63, 64, 65, 70, 71, 72, 80, 82, 83, 84, 107, 108, 117, 118, 127, 129, 130, 131, 132, 133, 144, 145, 146, 174, 179, 181, 195, 196
mandates, vii, 25, 26
manufacturer, 150, 164
mapping, 15, 79
marine environment, 15, 16
market, ix, 62, 85, 86, 197
market access, 62
market economy, 86, 197
marketing, 33, 60, 62, 82, 84, 120, 145
markets, 129
marriage, 28
mask, 20
Matrices, 76, 78
matrix, 2, 17, 23, 65, 66, 76, 78, 185
meanings, 47, 69
measurement, 62, 64, 70, 119, 194, 197
measures, xii, 14, 15, 37, 181, 182, 186, 187, 189, 191, 193, 194, 199
meat, 122, 123
media, 15, 38, 162
median, 31
Mediterranean, 6, 159, 163
Mediterranean climate, 163
melt, 30
membership, 79
men, 73
metabolic, 150
metabolism, 150, 162, 167, 168
meteorological, 159
methane, 167
Methylobacterium, 152, 154
Mexico, 28, 179, 197
microbes, 157, 162, 167
microbial, xi, 147, 148, 150, 151, 156, 157, 158, 159, 161, 162, 163, 165, 166, 167, 168, 169, 170, 171
microbial cells, 150, 156, 162
microbial communities, xi, 148, 156, 157, 159, 161, 162, 163, 166, 169, 170, 171
microbial community, 166, 167, 170
microhabitats, xi, 161
microorganism, 153, 154
microorganisms, xi, 147, 148, 150, 151, 156, 157, 158, 159, 161, 162, 163, 164, 165, 166, 167, 168, 169, 170, 171
microscope, 37, 148
microscopy, 162
microtubes, 150, 164
middle class, 27
migration, 69, 86, 91, 92, 100, 123, 154
military, 69
mineralization, 167, 168
mines, 185
Ministry of Education, 158, 170
Ministry of Environment, 145
missions, 3
mixing, 94
mobility, 156
model system, 148, 156
models, 44, 59, 148, 184, 186, 188, 189, 190, 191, 192, 193, 195, 197, 198, 199
modernization, 133, 145
mole, 141
molecular biology, 91
molecular fingerprints, 151, 165
mollusks, 68
money, 197
Montana, 109
moral standards, 107
morbidity, 94
morning, 94
morphological, 95
morphology, 95
mortality, 36, 38, 93, 94, 95, 106, 114, 125
moths, 21
motion, 28
motivation, ix, x, 85, 117
mountains, 88, 90, 111, 112
movement, 91
multidimensional, 45
Myanmar, x, 131
Mycobacterium, 154

N

Namibia, 145
Nash, 26, 41
nation, 22, 26
national, viii, ix, xi, xii, 3, 6, 7, 14, 15, 16, 21, 23, 26, 27, 28, 29, 30, 38, 43, 44, 46, 47, 48, 49, 50, 52, 54, 55, 56, 57, 58, 61, 62, 63, 64, 65, 66, 67, 68, 69, 70, 71, 72, 73, 76, 80, 82, 83, 85, 86, 87, 94, 117, 118, 119, 120, 143, 144, 154, 163, 173, 174, 179, 181, 189, 197
National Park Service, vii, 25, 26, 27, 29, 38, 40, 41
national parks, viii, ix, xi, 3, 26, 27, 28, 38, 43, 44, 57, 58, 61, 83, 85, 86, 94, 118, 120, 173, 174, 179, 189
National parks, vii, 14, 44
National Research Council, 28, 29, 30, 41

Index

native plant, 27, 31
native species, 27, 32, 34, 38
natural, viii, ix, xi, 2, 7, 14, 15, 16, 18, 19, 26, 27, 28, 31, 32, 33, 34, 38, 39, 44, 45, 49, 57, 59, 61, 68, 69, 71, 72, 80, 83, 85, 86, 87, 93, 98, 100, 106, 107, 113, 115, 122, 123, 124, 132, 134, 140, 145, 150, 156, 158, 159, 161, 162, 163, 167, 169, 170, 171, 173, 174, 175, 176, 179, 197
natural environment, 150, 158, 161, 162, 169, 171
natural resource management, 83
natural resources, ix, 28, 38, 44, 45, 71, 72, 80, 85, 86, 122, 123, 134, 197
nature conservation, vii, 1, 3, 49, 86, 174
Nebraska, 41
negative consequences, 17
neglect, 130
negotiation, 13
nematode, 95, 98, 100, 101, 109, 115
nematodes, 94, 95, 96, 97, 98, 99, 100, 101, 102, 106, 108, 112, 114
Nepal, 62, 84
Netherlands, 197
network, vii, 1, 2, 3, 16, 23, 118, 126, 127, 128, 129, 143, 146
Nevada, 28
New Mexico, 28, 197
New World, 13
New York, 40, 83, 130
New Zealand, 58
NGO, 47, 48, 49, 50, 51, 52, 53, 54, 55, 56
NGOs, x, 132, 144
nitrate, 167
nitrogen, 167
nitrogen fixation, 167
nodules, 100
noise, 127
nongovernmental, viii, 43, 47, 57
nongovernmental organization (NGO), viii, 43, 47, 57
nonlinear, 187
non-native, 31
non-profit, 63
normal, 35, 192, 195
norms, 133
North America, 4, 28, 90, 174
North Carolina, 190, 193, 196
Northeast, 132, 139, 146
nucleic acid, 150, 156, 162, 164
nucleotide sequence, 111
nucleotides, 155
nucleus, 148
nutrient, xi, 100, 161
nutrient cycling, xi, 161
nutrients, 167

O

observations, 109, 135, 185, 187, 191

Ohio, 199
online, 41, 42, 179, 180, 197, 198
opportunity costs, 123
optical, 152
optimization, 48
Oregon, 31, 152
organic, 167, 168
Organic Act, 26, 32, 38
organic matter, 167
organization, viii, 35, 43, 47, 63, 112
organizations, viii, 44, 57, 174
originality, 87
out-of-pocket, 188
overpopulation, 124
overtime, 152
ownership, ix, 85, 86, 144
oxidation, 168

P

Pacific, 150
PAN, 49
paper, ix, 41, 42, 58, 59, 61, 63, 64, 68, 85, 117, 118, 145, 146, 179
paradoxical, 14, 19, 26
parameter, 67, 74, 82, 185, 187, 190, 191, 192, 194
parasite, 100, 101, 110
parasites, 93, 94, 95, 100, 101, 103, 104, 105, 108, 111, 114
parasitic infection, 109
parasitic worms, 109, 111
parenchyma, 98, 100
Paris, 21, 22, 109, 115
PARK, 25, 43, 61, 117, 161
partnerships, 130
pasture, 118
pastures, 88, 99
pathogens, 94, 106, 125
PCR, 151, 158, 164, 171
pedagogical, 19
perception, 19, 65, 76, 82
perceptions, 62, 82, 83, 84
performance, 59, 63, 74, 75, 76, 82
periodic, vii, xi, 33, 36, 147, 173
personal, 82
pests, 13
philosophy, 26
phosphorous, 167
photographs, 36
phylogenetic, 162
phylogeny, 158
physical environment, 121
physical factors, 98, 113
physiological, 148
pilot study, x, 132
planning, 15, 20, 30, 35, 44, 45, 57, 58, 60, 142, 143, 144, 179, 198

plants, vii, 17, 20, 21, 22, 31, 92, 99, 139, 140, 146, 147, 148, 149, 159, 162, 163, 165, 167, 169
play, x, xi, 2, 14, 17, 28, 92, 93, 94, 100, 132, 147
Pleistocene, 89
pleura, 100
Pliocene, 89
pneumonia, 107
Poisson, 184, 185, 186, 191, 192
Poisson distribution, 184, 185, 186, 191
Poland, 90
politicians, 13
politics, 14, 49
pollen, 156
pollination, 17
pollutants, 169, 171
pollution, 94, 127, 169, 171
pond, 165, 167
poor, 95, 126, 144, 145, 169
POPs, 170
population, ix, x, 17, 18, 19, 21, 28, 34, 36, 37, 38, 69, 70, 85, 86, 88, 93, 94, 97, 99, 101, 106, 107, 108, 109, 110, 112, 113, 117, 118, 122, 127, 131, 133, 148, 170, 193, 194
population density, 19
population growth, 93
population size, 17
positive correlation, 82
positive relation, 79
positive relationship, 79
posture, 176
poverty, 16, 49, 55, 93
power, 28, 31, 32, 33, 35, 36, 38, 39, 93
power generation, 28, 32, 33
powers, 92, 139
predators, 2, 17, 92, 97, 113, 169
predictability, 30
pre-existing, 32, 33
preference, viii, ix, 33, 43, 48, 51, 52, 54, 55, 57, 58, 59, 60, 61, 63, 64, 65, 74, 76, 80, 182, 197
pressure, x, 19, 31, 93, 94, 118, 124, 132, 144
prevention, 72, 178
prices, 35
primary school, 49
priorities, 22, 44, 45, 52
pristine, 69, 169
private, 29, 50, 188
private sector, 50
probability, 30, 182
probe, 33
production, 17, 33, 55, 168
productivity, 95, 115
program, 33, 48, 178
prokaryotic, 148, 158, 170
prokaryotic cell, 148
promote, 26, 27
property, 135, 185
prophylaxis, 113
protected area, viii, x, xi, 3, 6, 13, 14, 15, 16, 17, 21, 22, 23, 44, 45, 49, 55, 57, 61, 62, 65, 83, 84, 86, 87, 117, 118, 119, 120, 121, 122, 123, 126, 129, 130, 132, 143, 144, 146, 173, 174, 179, 198
protected areas, viii, x, xi, 3, 6, 13, 14, 15, 17, 21, 22, 44, 45, 49, 55, 61, 62, 86, 87, 117, 118, 119, 120, 121, 122, 123, 126, 130, 132, 143, 146, 173, 174, 179, 198
protection, vii, viii, ix, 1, 13, 14, 15, 18, 19, 31, 32, 44, 45, 57, 58, 61, 68, 85, 86, 91, 94, 95, 107, 109, 111, 112, 113, 114, 145, 147, 163, 179
Proteobacteria, 155, 156
protocol, 150
protozoa, 148
proxy, 18, 182, 188, 189, 191
pseudo, 192
Pseudomonas, 153, 168
psychology, 83
public, viii, 14, 16, 18, 19, 20, 22, 28, 30, 33, 39, 43, 44, 47, 48, 49, 50, 52, 54, 55, 56, 181
public health, 22
public opinion, 14
public support, 20

Q

quality of service, 55
Quercus, 141
questionnaire, 47, 70, 135, 184, 188, 190
questionnaires, 184
quota sampling, 70

R

radiation, 98, 99
radical, 163
rain, 16, 88
rain forest, 16
rainfall, 115, 163
rainforest, 58
random, 47, 56, 185, 186, 199
range, 33, 36, 37, 39, 47, 87, 88, 95, 136, 137, 138, 144, 152, 153, 163, 167, 174, 175
rating scale, 72
ratings, 51
reagents, 150
reasoning, 2, 20
recall, 188, 189
recession, 87
Reclamation, 33, 35
recognition, 74
recovery, 32
recreation, 13, 20, 27, 29, 33, 38, 39, 57, 59, 83, 181, 182, 183, 184, 193, 194, 195, 196, 197, 198, 199
recreational, vii, xii, 25, 26, 27, 29, 30, 33, 44, 46, 57, 58, 144, 181, 182, 183, 184, 185, 189, 193, 195, 196, 197, 198, 199
recreational areas, xii, 181, 195
recruiting, 38

reduction, vii, 30, 31, 128, 145, 147, 168
reef, 15, 16
reefs, 15, 23
refuge, ix, 117, 197
regeneration, 22, 36, 38, 107, 125, 127
regional, 29, 144, 175
regression, 14, 48, 51, 185, 186, 187, 196, 198, 199
regressions, 186
regular, 92
regulation, 17, 31, 36, 72, 94
regulations, viii, 36, 61, 63, 71, 72
rehabilitation, 46
relationship, 65, 66, 70, 76, 77, 78, 79, 91, 92, 93, 167, 171, 182
relationships, 65, 66, 72, 82, 112
relatives, 152, 183
relaxation, 72
reliability, 68
religion, 140
religious belief, 132, 133
religious beliefs, 132, 133
remote sensing, 41
reproduction, 97, 100, 106
reptile, 178
reptiles, 68, 175
reputation, 26
research, vii, viii, 21, 25, 27, 32, 33, 34, 35, 36, 38, 55, 58, 59, 62, 63, 65, 82, 95, 108, 127, 145, 158, 170, 174, 195
researchers, 70, 124
resentment, 124
reservation, 100
reserves, 13, 23, 87, 130
reservoir, 30, 36, 41, 88, 109
reservoirs, 28, 29
residential, 199
resistance, 97, 98, 99, 100, 113, 175
resource management, 146
resources, ix, x, 23, 26, 27, 28, 34, 38, 44, 45, 59, 71, 72, 80, 82, 84, 85, 86, 107, 122, 123, 132, 134, 144, 169, 183, 196, 197, 198
respiratory, 95, 108
response time, 58
responsibilities, 55
responsiveness, 169
retardation, 95
revenue, ix, x, 117, 118, 120, 126, 129, 144
reverse transcriptase, 151
ribosomal, 150
ribosomal RNA, 150
rice, 135
rings, 36, 37
risk, 22, 35, 36, 60, 62, 157
risks, 120, 169
rivers, 19, 34, 134, 149, 196
RNA, 150, 151, 152, 153, 154, 155, 156
Roads, 127
robustness, 68
rocky, 91

rodents, 175
Romania, 89, 101, 104
routines, 187
Royal Society, 21, 22
rubber, 29
ruminant, 106, 109
rural, 18, 23, 49, 55, 57, 83, 144
rural development, 23, 57, 144
rural poverty, 49, 55
Russia, xi, 147, 148, 149, 150

S

sacred, x, 69, 132, 133, 134, 135, 136, 137, 138, 139, 140, 141, 142, 143, 144, 145, 146
sacrifice, 126, 139, 140
safeguard, 38
safeguards, 121
safety, 30, 91, 111
salinity, 148
salmon, 196
Salmonella, 106, 110
salt, 31
saltwater, 196
sample, 37, 70, 73, 150, 152, 156, 166, 182, 190, 194, 195, 196, 199
sample mean, 195
sampling, viii, xii, 43, 47, 56, 70, 181, 184, 190, 193, 198, 199
sanctions, 132
sand, 34
Sarajevo, 108
SAS, 113
satisfaction, 66, 67, 76, 79, 80, 82
scarcity, 20, 23
Schmid, 114
school, 49, 50
scientists, 93, 148, 162
scores, 120, 121, 122
sea level, 87, 88, 90, 99, 100
search, 69, 91, 92, 99, 148, 158, 165, 170, 179
searching, 91, 92
seasonal variations, 163
secondary data, 197
security, 120
sediment, 28, 31, 34, 35, 39, 165, 166, 167
sedimentation, 41
sediments, 28, 38, 164
seed, 31
segmentation, 91
selecting, 51, 82
senescence, 124
sensing, 41
series, 33, 35, 184, 193
services, x, 17, 22, 45, 55, 62, 132, 140, 144, 182, 183
settlements, 46, 69, 133
severity, 174

sex, 93
sex ratio, 93
shade, 92, 100
shape, 87, 90, 100, 174
sharing, viii, x, 2, 43, 47, 50, 51, 52, 54, 55, 56, 131, 190, 191
shear, 130
sheep, 114
shelter, 92
Shigella, 153
shores, 163, 165, 167
shrubs, 140, 141
sign, 192, 193
signs, 189
similarity, 99, 153, 154
simple random sampling, 56
simulation, 34, 35
simulations, 33
singular, xi, 161, 167, 175
sites, vii, xi, 1, 14, 30, 37, 38, 68, 69, 87, 90, 94, 98, 99, 147, 148, 149, 150, 151, 152, 155, 156, 161, 163, 165, 167, 169, 181, 182, 183, 184, 187, 189, 193, 196
skeleton, 91
skewness, 184
Slovakia, ix, 85, 86, 87, 89, 90, 95, 100, 104, 105, 106, 110, 113, 114, 115
Slovenia, 90, 107
Smithsonian, 110
Smithsonian Institution, 110
snakes, 175
social activities, 69
social group, 47
social life, x, 131
Social Services, 199
social stress, 94
social welfare, 197
socioeconomic, 144, 182
software, 165, 187, 198
soil, 62, 87, 92, 132, 140, 171
soils, 88
solar, 98, 99
solutions, 48
songbirds, 92
sovereignty, 62
Spain, xi, 59, 109, 147, 148, 149, 161, 163, 164, 171
spatial, xi, 2, 15, 16, 17, 18, 161
spawning, 31, 33, 34
species, vii, ix, x, xi, 1, 2, 3, 6, 7, 13, 14, 15, 16, 17, 18, 19, 20, 21, 22, 23, 27, 28, 30, 31, 32, 33, 34, 36, 38, 49, 68, 85, 86, 88, 89, 90, 92, 93, 95, 96, 97, 98, 99, 100, 101, 106, 107, 108, 109, 110, 118, 120, 124, 125, 126, 127, 129, 131, 132, 135, 140, 141, 143, 145, 169, 170, 171, 173, 174, 175, 176, 178, 182, 183, 195
species richness, 2, 18, 22, 145
specificity, 109
specter, 26
spectrum, 33

speculation, 162
speed, 189
spheres, 174
spine, 96
spiritual, 132, 135
springs, 86, 119
SPSS, 48
SSI, 47, 60
stability, xi, 161, 169, 170
stages, 97, 98, 99, 101, 109, 113, 133, 155
stakeholder, 33, 55
stakeholders, 33, 144
standard error, 185
standards, 107
starvation, 93
statistical analysis, 48
statistics, 2
statutes, 31
stochastic, 2
stomach, 93
storage, 28
storms, 98, 145
strain, 155
strains, 151, 167
strategic, 82, 175, 179
strategies, viii, 43, 44, 45, 47, 48, 50, 51, 55, 56, 82, 94, 106, 145, 146, 148, 157, 170
stratification, xii, 181, 182, 186, 187, 190, 192, 193, 195, 197, 198, 199
streams, 34, 134, 140
strength, 169
stress, 17, 37, 94, 126
structuring, 162
subgroups, x, 131, 133
subjective, 19, 45
subsistence, 144
substances, 97
substitutes, 187
substrates, 41, 100, 162, 167, 168
suburbs, 18
suffering, 100
Sulfide, 168
sulphur, 167
summer, 33, 35, 36, 91, 92, 94, 98, 159
supernatural, 140
supply, 28
suppression, 36, 37, 38, 125, 174, 175, 179
surplus, xii, 181, 182, 183, 191, 193, 194, 195, 196
survival, 142, 155
sustainability, 44, 54, 59
sustainable development, 84
sustainable tourism, 71, 117
swamps, x, 117, 118, 124, 125, 126
Switzerland, 22, 23, 101, 103, 106, 109, 129, 130
symbols, 76, 78
synchronization, 44
synthesis, 79, 82
Syria, 115
systems, 15, 39, 49, 143, 148, 156, 157, 169

T

taiga, 90
Taiwan, viii, ix, 61, 62, 63, 68, 70, 83
tangible, 34
Tanzania, 3, 123
targets, 33
taxa, 18, 21
taxonomic, 2, 110, 152, 175
taxonomy, 158
teaching, 19
technology, 16, 28, 29
temperature, xi, 30, 33, 39, 87, 88, 91, 92, 98, 100, 147, 148, 149, 150, 151, 152, 153, 155, 156, 157, 158, 164
temporal, 16
Tennessee, 31
Tennessee Valley Authority, 31
tension, 26
tenure, 129, 130
territorial, 86, 94, 140
territory, ix, 6, 13, 14, 26, 70, 85, 86, 88, 90, 91, 92, 93, 94, 95, 100, 101, 106
testimony, 26
TGA, 151, 164
Thailand, 84
thawing, 91
Theodore Roosevelt, 26, 41
theory, 58, 83, 193
therapy, 113
thermal denaturation, 158
thermophiles, 148, 150, 156, 159
Thomson, 95, 115, 124, 125
threat, xi, 26, 128, 173, 174, 175, 178
threatened, vii, ix, x, 1, 2, 3, 14, 19, 31, 85, 86, 110, 122, 123, 131, 132, 140, 171, 174, 175
threatening, 126, 140
threats, 17, 30, 48, 122, 123, 124, 129, 176
threshold, 169
thresholds, 92
Tibet, x, 131, 134, 139
timber, 145
time, 19, 20, 26, 39, 44, 45, 47, 89, 90, 92, 95, 97, 99, 106, 112, 148, 151, 152, 155, 156, 174, 175, 179, 182, 187, 189, 193, 195, 196, 197, 199
time frame, 151, 155
time use, 45
timing, 35
tissue, 95, 98, 100
tolerance, 124
tourism, vii, viii, ix, x, 1, 2, 16, 18, 19, 45, 49, 57, 58, 59, 61, 62, 63, 64, 65, 69, 71, 72, 83, 84, 86, 91, 94, 117, 118, 120, 121, 122, 123, 124, 126, 127, 128, 129, 130, 132, 144, 145
tourist, x, 19, 57, 62, 69, 71, 72, 99, 117, 118, 119, 120, 121, 122, 126
toxic, 92, 97
toxicology, 27
trachea, 98
trade, 32, 58, 59
trade-off, 182
tradition, 71, 72, 134
traffic, 29, 127, 128, 199
training, 145
traits, 20, 21
transcriptase, 151
transcription, 151
transfer, 55
transformation, x, 118, 132, 169
transformations, 157, 161, 167
transition, 82, 100
translation, 4, 64
transmission, 109
transportation, 30, 150, 164, 182, 188, 197
traps, 34
travel, xii, 181, 182, 184, 186, 187, 188, 189, 190, 191, 192, 193, 196, 197, 198, 199
travel time, 187, 189, 196, 199
treaties, 106
trees, 37, 38, 132, 139, 140, 141
trend, 18, 27, 29, 106, 169
Triassic, 88
tribal, x, 33, 70, 131, 132, 133
tribes, x, 132, 133, 134, 139
trout, viii, 25, 29, 31, 34
Tsuga, 142
tundra, 90
Turkey, viii, 43, 44, 45, 46, 49, 57, 58, 59, 60, 197

U

U.S. Geological Survey, 40, 42
unemployment, 29
UNESCO, 3, 87, 183
unification, 4
United Nations, 23, 49, 58, 60, 106, 173
United Nations Development Program, 46, 49, 58, 60
United States, 4, 28, 34, 42, 44, 199
universe, 66
upload, 41
urban centers, 195
urbanization, 69
urbanized, 19, 142
Utah, 28, 30, 197

V

validity, 45
values, viii, 20, 21, 25, 26, 28, 30, 31, 32, 34, 35, 43, 44, 45, 51, 52, 57, 58, 59, 62, 65, 72, 74, 76, 134, 152, 155, 157, 183, 184, 187, 191, 193, 194, 196, 197, 198
variability, 17, 157, 174, 175

variable, xii, 31, 48, 51, 62, 115, 163, 181, 182, 184, 186, 187, 188, 189, 190, 191, 192, 193, 195
variables, 44, 51, 71, 79, 182, 185, 188, 190, 191, 194
variance, 184, 185
variation, 21, 31, 59, 63, 106, 157, 174, 187
vector, 65
vegetation, viii, 25, 35, 36, 40, 41, 42, 87, 88, 99, 127, 132, 135, 140, 145, 175, 178
vehicles, 121, 127, 128
venue, 84
Venus, 22
vertebrates, 175, 176
village, 135, 138, 140
Villagers, 47, 50
visible, 15, 149, 162, 163, 167, 169, 182, 195
visual environment, 169
vulnerability, 19, 26, 30, 176

W

wage rate, 189
wages, 182
walking, 144
wastewater, 159
water, x, 23, 28, 29, 30, 31, 32, 33, 34, 35, 39, 40, 62, 69, 92, 98, 99, 117, 118, 124, 125, 148, 149, 150, 163, 165, 167, 171, 172, 196, 199
water table, 124, 163
waterfowl, 58
weakness, 48, 82
welfare, 33, 140, 182, 186, 187, 193, 194, 199
wellbeing, 20, 27
Western Europe, 89
wetlands, 17, 174

wilderness, vii, 1, 2, 4, 13, 26, 30, 183
wildfire, 174, 175, 176, 178
wildfires, xi, 173, 175, 176, 179, 180
wildland, 180
wildlife, ix, x, xii, 21, 26, 33, 45, 49, 56, 62, 83, 84, 117, 118, 120, 121, 122, 123, 124, 126, 127, 129, 130, 144, 176, 179, 181, 182, 183, 189, 190, 195, 196
wildlife conservation, x, 117, 124
wind, 92, 149, 156
winter, 36, 91, 93, 94, 98, 115
wisdom, 14
withdrawal, 33
women, 73
wood, 49, 55, 59, 145
wood waste, 59
woods, 16
workers, 127, 132
World Bank, 84, 146
World War, 93
World War I, 93
World War II, 93
worm, 22, 109
worms, 101, 109, 111, 114

X

xenobiotic, 171

Y

Yellowstone National Park, xi, 147, 148, 149
Yugoslavia, 101